THE CONQUEST OF WATER

THE CONQUEST OF WATER

The Advent of Health in the Industrial Age

Jean-Pierre Goubert

Introduction by Emmanuel Le Roy Ladurie

Translated by Andrew Wilson

Princeton University Press
Princeton, New Jersey

© 1986 by Editions Robert Laffont
This translation © 1989 by Polity Press

Published by Princeton University Press
41 William Street, Princeton, New Jersey 08540

Library of Congress Cataloging-in-Publication Data

Goubert, Jean-Pierre
[Conquête de l'eau. English]
The conquest of water: the advent of health in the Industrial Age.
Jean-Pierre Goubert : introduction by Emmanuel Le Roy Ladurie :
translated by Andrew Wilson.
p. cm.
Translation of: La conquête de l'eau.
Bibliography: p.
Includes index.
ISBN 0–691–08544–7
1. Water–Purification–History.   I. Title
RA591.G6813   1989   89–8440
    628.1′62′09–dc20   CIP

Printed in Great Britain

Contents

Introduction

The preoccupation with water is, according to Jean-Pierre Goubert, one of the subdivisions of the religion of progress. This particular creed has its priests (doctors, architects, engineers), its temples and altars (aqueducts, baths. . .) and its congregations. An early document on the subject appears in the 'medical topographies' of the eighteenth and early nineteenth centuries. It is a distant harbinger of our present ecology. The topographies had their origin in enquiries carried out in the eighteenth century by the Royal Society of Medicine. One of these enquiries was carried out by a Dr Berthelet who, in Lyons in 1783, denounced the bad drainage; he wanted to drive away miasmas and increase the number of fountains. These were the concerns of the old-fashioned 'meteorological' school of medicine which saw air and water as carriers of disease. This school also claimed, rightly or wrongly, to be geological; one doctor in Thiers spoke of the silt-laden water, while in Rennes in 1789 Salmon described the water as sulphorous and hard.

These doctors were interested of course in the way in which water was drawn off, by porters, through wells and fountains, etc. The doctors of the time developed the notion of a link between a particular concept of water technology, in this case wells, and various diseases (congestion, obstructions of the viscera and the lower abdomen, swelling of the hypochondrium, cachectic diseases from which young nuns perished, etc.). The work of Daniel Roche is useful on this point; in 1787, the Academy of Sciences, which was granted a charter by the King, issued advice on the need to install drains and sewers in hospitals; the members of the academy were thus fulfilling the role left vacant, though not for long, by professionals and administrators. It is true that there was no question, as the *ancien régime* was drawing to a close, of installing a private water supply, toilet and bathroom in every dwelling. As early as the end of the eighteenth century, however, the solutions that were universally adopted in the nineteenth century were being put forward.

Large cities such as Paris were to have pumping centres, or at least water would be supplied to the capital by gravity, as indeed it was at the beginning of the nineteenth century as a result of the diversion of the river Ourcq.

Water retained its sacred, purifying role for a long time, whether this was reflected in the mineral springs honoured even in protohistorical times by *ex-votos* which had nothing Christian about them (and for a good reason), or quite simply in the holy water at the entrance to our churches. Was it really Lavoisier who, in the eighteenth century, gave a decisive push to the secularization and demystification of water by analysing it and showing that it could be broken down into hydrogen and oxygen, thus disproving the notion that it was an element like air, earth and fire. . .? In any case, mystery gave way to science, religion to technology and salvation to health.

In discussing the social uses of water, Jean-Pierre Goubert makes the necessary distinction between town and country. The prefect of a department in central France was not concerned by the fact that peasants were drinking water that might be stagnant; on the other hand, he found it quite normal that in 1771 the city authorities in Bourg had built fountains and drained the ditches that were a breeding ground for malaria. The enormous problems created by the danger of malaria should be mentioned in this respect: at the beginning of the nineteenth century, many areas of stagnant water, such as the Dombes and the ponds of the Languedoc, were so badly affected by the disease that the peasant population was seriously debilitated.

After the medical topographies a new source of information emerged during the First Empire; these were the papers of the Consultative Committee on Arts and Manufactures attached to the Ministry of the Interior, which was to continue its work during the Restoration. For example, the Committee expressed its opinion in 1826 on the proposals of an individual who wanted to use powdered coal as a protection against the putrefaction of water in the holds of ships.

The ideological foundations of these various initiatives had hardly changed; it was still a question of 'taking care of water', in other words of Nature, and of preventing disease by attacking it downstream of the point of infection, for want of the ability to cure it at source. In the absence of any effective treatment for diseased individuals, the wise thing to do was to ameliorate the environment. Interest centred less on man than on his habitat, in the general sense of the term.

It is true that governments of the period were not necessarily won over to the crusade for the purification of water. Witness the response of the Consultative Committee on the Arts to a supporter of purification: 'It would be desirable,' replied the Committee to this man of good will, 'if the number of water purifying establishments were to be increased, but the government will play no part in this. It is up to private individuals to set up these establishments; the most that the public authorities should do is to protect them.'

Having prepared this initial ground, Jean-Pierre Goubert tackles the subject with the aid of several large bodies of texts; since it was impossible to read everything, the author decided to concentrate on the *Bulletin of the Academy of Medicine* and the *Proceedings of the International Conferences on Hygiene*.

Water was not at the heart of the preoccupations of the Academy of Medicine between 1830 and 1940. Only 6 per cent of its meetings were devoted to this 'precious liquid', while three-quarters of them were given over to diseases (cholera, typhoid and malaria) and less than one-quarter to hygiene in the strict sense. Of the three diseases listed above, typhoid is the most frequently mentioned, having been discussed at 40 per cent of the meetings. Thus the principal concerns were, predictably, strictly medical. However, concern with hygiene itself did begin to emerge from 1891 onwards. This change, that took place in the final decade of the nineteenth century, seems to be linked to the emergence of the Pasteurian mentality and approach to medicine. In any event, the papers presented at the Academy are extremely valuable from the point of view of social and even geographical history; a paper given by a doctor from the Nièvre district mentions the market town of La Machine in 1896: 'In the middle of the century, there was virtually no paving in the streets, and what there was was in very poor condition; there were no sewers, no refuse dump, and, with a few exceptions, no flushing lavatories; drinking water was supplied by a few old wells in the square. In contrast, in 1896, the streets were properly surfaced and had pavements and a sewerage system and public water supply were well established.'

The diseases caused by stagnant or putrefying water have a sometimes mysterious history. For example, how can the decline in malaria about 1885 be explained? The improved welfare of the population and the spread of primary education do not in themselves provide an adequate explanation. On the other hand, as far as cholera and, particularly, typhoid, were concerned, the influence of Pasteur that began to be felt after 1890 sparked off a real debate on the pollution of water by microbes. Herein lies the radical change. Since Lavoisier, the old mythology of water had been called into question; nevertheless, the classic distinction between stagnant water and running water still persisted, with the latter being considered healthier. Under Pasteur's influence, this criterion lost some of its importance; the notion of bacteriologically pure water became fundamental, at the same time as a new obsession with pure water (this time 'scientifically' based) was spreading in certain quarters, with people constantly washing their hands and refusing to drink water unless it had been filtered or boiled.

Certain standards became established during this period; demands were made for a supply of 100 to 120 litres of water per person per day, an optimal supply in France, but which was far from being attained at the end of the nineteenth century. The sanitary state of some towns was still giving rise to horrified statements, to the extent that references were made to the 'fecal peril'. In Beauvais, the number of watertight cesspits (about 350) and mobile lavatories (570) proved to be too low for a town of approximately 15,000 inhabitants. As for sewage disposal, the situation was hardly any better. The problem of 'washing the water', to quote a picturesque phrase, still remained.

In this area, the International Conferences on Hygiene (1852–1908) provided a wealth of relevant information, and some symbolic or even effective measures were taken. In 1885, the procedures for the bacteriological analysis of water were laid down; as far as sanitary equipment was concerned, the English model was put forward as being more hygienic than the French systems of the period (in this respect, see Frédéric Amiel's observations on the deplorable dirtiness of the French under the Second Empire). Many real changes were to be made, however, and not only in the deliberations of those attending the conferences. The typhoid epidemic of 1892 in Paris gave rise to the first systematic sampling of water for laboratory analysis. The myths relating to some water formerly considered to be pure were destroyed once and for all, and the idea began to gain ground that there should be a compulsory supply of water that was universal in the same way that education and even military service were for young people. This new type of water supply required colossal financing at the level of the tens of thousands of parishes that existed in France. It led to increased state intervention and was a forerunner of the sometimes directive modernization with which we have been well acquainted for a long time. From this point of view, the Public Hygiene Act of 1902 emanated from the radical lay, even masonic, elites that were assuming responsibility for the destiny of France.

If the water supply was in some instances nationalized, it was also being industrialized. The supporters of both developments had scant regard for deeply ingrained customs; a change was taking place from habits similar to those that ethnographers were still observing outside Europe to the considerable hygiene of our time. From time to time in the course of Jean-Pierre Goubert's narrative, the dynamic specialists of the discipline emerge as they flourished almost a century ago. Durand-Claye was the successor to Belgrand in Paris; in one year, he wrote fourteen articles on hygiene standards with respect to lavatories, and three more on mobile latrines.

It is at this point that a new area of Goubert's research emerges: the doctoral theses of the Faculty of Medicine in Paris. Unusually for the period in question, these theses included some devoted to various aspects of water, although they accounted for only 1 per cent of the total submitted. And of this meagre collection, 80 per cent dealt with the curative treatment of typhoid, cholera and malaria. Here again, the Pasteurian revolution of the years between 1880 and 1890 was decisive, in view of the fact that the medical thesis of the time was not, with some exceptions, the result of research activity. This 'literary genre' tended rather to follow in the wake of discoveries and was part of the process of popularization intended to make these discoveries better known; they had in fact taken place independently of the enquiries of the young thesis writers. To go into details, the role of vegetables contaminated by sewage and of polluted shellfish is noted in the post-Pasteurian theses relating to typhoid. The silences, however, are more eloquent. The theses generally have nothing to say about the countryside and the rural world; the rural exodus and the consequent depopulation of the countryside were allowed to take place without any attempt at the time to reform village life: improvement was to come later. There was also silence on the cost of a water supply system.

Among these works written by future doctors there was *one* study of the pollution of water, flora and fauna. This was, however, a truly exceptional piece of research, unique in its genre. In general, a distinction has to be made between the young medicos who, statistically, had little interest in water and, on the other hand, the somewhat older doctors of medicine, established men who had submitted their theses a long time before. They were now elected representatives of their local communities or persons of influence at department level and were beginning to play a part in major public works, including water supply schemes.

Among the sample chosen by Goubert, the first thesis on the need to wash dirt off children was published in 1843. It was still only a question of washing the decent parts of the body, but the author of this work, a certain Doctor Ruelle, should be acknowledged in passing. On the other hand, it should be noted that, even in a thesis written in the 1930s, the traditional 'atmospheric' notion of humidity was in conflict with the infinitely more up-to-date Pasteurian medicine.

A current of optimism runs through this collection of theses. The concepts of cleanliness and purity appear in them, together with references to hygiene, asepsis and prophylaxis. Some of the thesis writers were, in their own way, historians; men of the left, they compared the seventeenth and eighteenth centuries unfavourably (and not always unjustly) with the purifying 'enlightenment' of the Revolution and the nineteenth century; curiously, this unfavourable comparison also held good for the Middle Ages, which they considered, rightly or wrongly, to be hygienic and clean.

The vogue for hydrotherapy was echoed in the improvements in hydrology which were a particular feature of the 1860s. It is true that the 1880s saw the diffusion of Pasteur's ideas, but it was in the ten years after 1860 that the foundations were laid; with Pasteur's assistance, this decade finally saw the demise of Pouchet's theories on spontaneous generation. Thus, *ipso facto*, the process of pasteurization, or heating of liquids, that several great civilizations had already practised in all innocence (the Chinese had for a long time boiled the water used to make tea), became indispensable. This same period was important in several other respects: the optimum volume of water 'per head and per day' rose from 100 litres around 1860 to 1 cubic metre in 1880; this was doubtless a purely theoretical figure, and one which applied more to England or the United States than to France, a distinctly more niggardly country. At the same time, as a direct result of the diffusion of Pasteur's discoveries, suspicion of wells and water tanks was growing. Finally, some things that we take for granted, such as washbasins, if not the WC, assumed their definitive shape during this period: the taps on washbasins were now placed above rather than below the basin, and so on.

It is a truism that the great epidemics occurred well before these years. The cholera epidemic of 1832 had several consequences in this respect: as a result of the catastrophe, medical committees composed of public-spirited people were set up in each department, a move which had been felt for a long time to be desirable. The classic measures in the fight against infection (removing cemeteries and dunghills from close proximity to wells, springs, etc.) had been implemented since the eighteenth century, particularly in the case of graveyards. From Goubert's point of view, the intervention of the mayors was more central; after 1832 some of them developed an interest in hygiene, sewage disposal, cesspits and wells. Unfortunately the measures taken by a town councillor in Normandy against water used for washing, which was suspected of transmitting the cholera bacterium, appear to have been more symbolic than effective.

Among the many items of legislation introduced from the 1850s onwards, the laws of May 1922 which laid down standards are of particular interest; they justified the monitoring of water in the name of 'the patriotic struggle against the depopulation of the French nation'. A clarion call indeed! Nevertheless, an adequate legislative framework was emerging, as Parliament became increasingly interested in the number of annual analyses of water purity relative to the number of inhabitants in a town, in thousands.

The issue of the 'sanitary record' was also raised; it had already been highlighted by an earlier publication, the *Annuaire des eaux*, or *Water Directory* of 1851. This publication was centred on the Paris basin as a result of the pressures exerted by the inevitable centralization. This concentration on the Parisian system, and thus on our capital, underlines, in passing, the evolution of a Rousseauesque culture; henceforward, the city was no longer a centre of beauty and civilization (the idealization of the city, which dated from the sixteenth century, had long since begun to pall); on the contrary, the town was presented as a pollutant which contaminated the river downstream. Nevertheless, the old stereotypes refused to die; for some time in the nineteenth century, the laxative properties of the water of the Seine continued (rightly!) to be praised.

One archaism was even revived; the status of the diviner, that water sorcerer referred to as early as the German Renaissance, was enhanced in the nineteenth century by the magic of electricity.

Having given prominence to the new knowledge that laid the foundations for the conquest of water, Jean-Pierre Goubert, in the second part of his book, investigates the means by which this conquest was achieved, and in particular the modes of social diffusion via the press, schools and hospitals.

The author has scrutinized the press in order to reveal the successive cultural images of water. To begin with, Jean-Pierre Goubert gives proof of the iconographic sources that he found in various periodicals. For this purpose, he has chosen *L'Illustration*, which was read mainly by the middle classes, and *Le Petit Journal* which, according to Zola, was read eagerly by *concierges*, the working classes and people of modest means.

The statistical analysis based on these two periodicals leads to some interesting discoveries: to judge from the advertising, the working-class readers of the *Petit Journal* were concerned with their hair (threatened, it is true, by baldness) and their teeth, which were in a deplorable state. On the other hand, the readers of *L'Illustration* were interested in all aspects of health and hygiene, starting with beards, a luxury item to which more advertising space was devoted than anything else. It goes without saying that, as proved by the illustrated press cuttings, the upper classes washed more! With respect to drinking water (considered in the theories of ancient medicine to be a means of internal 'cleansing'), the two publications converge. In general, references to prices are more frequent in the working-class journal (44 per cent) than in the so-called middle-class publication (26 per cent); this difference should be viewed against the background of the widely differing circumstances of the readerships of the two periodicals. The publication aimed at the more affluent readership made greater use of images, particularly photographs; this was not to change until the present century. *Le Petit Journal* used the printed word and drawings.

The masses had a fondness for mineral waters and even soap, whereas the elite, more intent on the pursuit of luxury, went to Evian or Vichy.

Scientific arguments were not much deployed in the columns of these publications. The elitist press emphasized the beauty of the body, aesthetic themes and others related to pleasure. This selective 'sales gambit' was subsequently to lead to the mass culture of our century, which was in its turn to be disparaged by the snobbishness of the elites of the 1960s and 1970s! The 1914–18 war did little to encourage advertising of this kind, although the greater hedonism of the interwar period led to a sharp increase in sales arguments based on the pleasure principle.

Between 1880 and 1920, there was a cyclical increase in advertising relating to water; there was a corresponding fall between 1929 and 1937–43, in response to the economic crisis and war that followed the Roaring Twenties. In all this, rationality was only a minor concern. In this press file, hydrotherapy, which is relatively ineffective, plays a much greater role than sanitary equipment, which was nevertheless to be much more useful in raising public health standards. Pseudoscientific testimonials, when they existed at all, were initially used to promote spas, despite the fact that their therapeutic value proved to be slight in the extreme. This reflects the continued attraction of a medical past based on Gallic, Gallo-Roman and medieval traditions, in which the healing properties of springs played a major part.

From the point of view of advertising, the age of the child as consumer had not yet arrived. More children were born then than now, but they did not feature so strongly in the advertising of the day. When it came to heads, toothless mouths did not feature at all in the press, although baldness, as an entirely honourable sign of virility, was often depicted. These advertisements depicted mineral water as a panacea rather than just a specific remedy.

The public education system was another route for the diffusion of hygiene. Jean-Pierre Goubert begins his examination of the education system with school buildings, which were gradually brought up to the standards laid down by statute in 1880. A short digression on the need for air and light in classrooms brings our author once more to the question of lavatories which were to be purified by water as a matter of priority. As early as the ninth century, the monks of the Abbey of Saint-Gall had individual lavatory seats. By the end of the nineteenth century, schools had not always reached this optimal level; it goes without saying that the doors of the WCs had a gap at the top and the bottom, in order to facilitate surveillance of pupils' morals. Anxiety about the water-related aspects of pupils' health emerged as early as 1848, when the Academy of Medicine expressed concern about the ravages of typhoid among secondary and boarding-school pupils.

Despite a tentative beginning in about 1835, the primary school medical inspectorate did not finally come into being until 1879.

Initially, only a minority of departments conformed to legal regulations in this respect. It will come as no surprise that the department of the Seine led the entire country in the area of dental, eye and ear as well as general medical examinations. National legislation and statutes in this area date from 1886 and 1887 and began the process by which schools were medicalized. In 1882, in a symbolic exchange, Ferry made hygiene part of the school curriculum, while at the same time taking some aspects of religious instruction out of it. Were the two really mutually exclusive? This question could be debated almost ad infinitum, although reflection would seem to suggest a negative answer.

Schoolteachers, in their turn, carried out 'cleanliness checks'. The celebrated *Tour de France par deux enfants* suggested that the body could be washed no matter where one happened to be, because, in the author's inspired words, 'there is no shortage of water in France.'

There remained what Jean-Pierre Goubert calls the 'parallel school', in which politicians, doctors and hygienists were active. In this respect, the role of Clémenceau, a descendant of a long line of doctors, should be stressed; the future 'Tiger' had a good understanding of the social problems of the working classes, as revealed by his intervention in 1904 on the dangers of white lead. The German example should also be highlighted; in the name of the advancement of hygiene, truly palatial public baths sprang up in large towns and cities. In France, however, a certain degree of resistance on the part of the working classes should be noted; trade unionists were unwilling to allow worker representatives to be in charge of hygiene regulations in firms because they would, according to the militants, be 'the accomplices or victims of the employers'. In this respect, mention should be made of the famous refusals by certain representatives of the world of work to proscribe alcohol during a meeting with Mendès-France more than thirty years ago. There then emerged the many booklets on hygiene which flourished from the *ancien régime* onwards. They purported to teach the public the basics of hygiene in twenty-five lessons, just as similar publications today claim to teach English in three and German in six months. They had rivals in the public health columns of the newspapers, the conferences on hygiene which were aimed at an educated audience and the reports directed at an increasing number of amateurs and all those who had passed the basic school certificate. Thus in the course of the nineteenth century a new relationship to water emerged, a relationship not entirely devoid of paternalism and scientism.

After the press and the schools, hospitals played an important role in the social diffusion of this conquest. They were places where the poor were gathered together and where water was conceived as one element of the collective hygiene that hospitals imposed in the name of hydrotherapy. Hospitals had long been concerned with water: at Saint-

Louis in Paris, the great reservoir was constructed in the seventeenth century; the paths taken by the pipes and the pump which conveyed the water to the various parts of the hospital buildings were perfected during the eighteenth century. At the Hôtel-Dieu in Paris, the benefits to be gained from the hygienic use of water had been realized since 1762. In about 1830, once again in Paris, some US students expressed their admiration of the water supply system in local hospitals. On the other hand, the hospital in Lyons was condemned for its lack of running water. Nevertheless, in the interests of equity, account should be taken of the difficult conditions of the period; taps, even when they existed, spat forth a brackish, foul-smelling, reddish liquid which it was virtually impossible to use, even for washing clothes and floors. Before 1884, only five Parisian hospitals had running water for domestic purposes, namely l'Hôtel-Dieu, Bichat, Saint-Louis, Tenon and the Maternity Hospital, where regular washing of hands by midwives had reduced the mortality rate among women in childbirth. For a long time, in fact, the isolation of the sick as a means of fighting contagious diseases took precedence over the supply of water. Despite these inevitable priorities, all hospital planners were at least theoretically concerned with the quality of water, as witness the projects realized by Petit, Poyer and Leroy in 1774, 1777 and 1786 respectively.

The third part of the book analyses the changes brought about by the conquest of water. The starting point is an obvious fact that is sometimes forgotten, namely that water, which was originally completely free of charge, became an industrial and commercial product, and sometimes even a profitable investment. Henceforth, the conquest of the noble liquid by industry and commerce occupies Goubert's thoughts, the starting point for which is the story of a new arrival in Paris in about 1790. He suffered from colic and diarrhoea because the water in the capital, according to him, was poorly filtered. At the end of the eighteenth century, people began to think about the water they drank, unlike their heedless forebears. From this point of view, the first event of importance was the first trial, in August 1781, of the steam pump developed by the Périer brothers. Cost price comparisons showed that this pump could supply water much more cheaply than the water carriers. The steam pumps in Paris were delivering 200 cubic metres of running water in 1788, which was not a vast amount; by the beginning of the nineteenth century, however, this figure had risen to between 6,000 and 8,000 cubic metres. The example had come from the English, who at the same time brought us steam engines and new ways of using water. The establishment of the Compagnie Générale des Eaux in 1853 marked the beginning of a more advanced phase of development, in which water would be supplied to city-dwellers and industrial establishments, but not yet to peasant farmers. The shareholders were drawn from the imperial elites of the reign of Napoleon II: aristocrats, financiers and Government Ministers. The Rothschild family played a major role in the financing of the company. Parisian business interests, aided by their counterparts in Lyons and London, were responsible for most of the initial investment. The minimum return guaranteed by the authorities was 5 per cent; the likely profit seems to have been around 25 per cent per annum. Large cities such as Paris, Nantes and Lyons headed the queue of applicants; in an area of staggering technical progress, the anticipated value added was enormous. The price of water was to be reduced four or fivefold in comparison with the past, as a result of the new methods being implemented by large companies. There was something in it for everybody, without it being necessary to moralize for or against the enormous profits that were made. The infrastructure remained the responsibility of municipal authorities, but the most profitable part of the venture fell into private hands; the development of the Côte d'Azur is reflected in the installation in the 1860s of water supplies and drainage systems in Nice and Villefranche-sur-Mer. In the 1870s and 1880s it was to be the turn of Monaco, Vence, Antibes and Hyères. In Brittany, the Emerald Coast benefited from the same injections of new money. The rate at which water supply networks were extended accelerated each decade between 1900 and 1940, with the exception of the Great War. Before the Second World War, the large companies, the Compagnie Générale and Compagnie Lyonnaise, supplied water to 50 per cent of town-dwellers. In the last twenty years of the Third Republic, small local companies were bought up and concentrated by the two large companies. However, the still decisive role of direct municipal management of the water supply in smaller towns should not be ignored. Subsidies from central Government and the departments made it easier for municipalities to embark on such major projects. Jean-Pierre Goubert notes, in passing, how difficult it is to conduct a serious analysis of the price of water, which depended in particular on the hydrographic situation in a particular area. It was often lower in mountainous regions and in well-watered plains; as in Roman times, it was higher in those areas where it was necessary to build aqueducts and reservoirs for a demanding clientele. The history and geography of needs also have a contribution to make; the tendencies towards intersocial imitation, dear to the heart of Maurice Agulhon, led to an ever-increasing demand for water among city-dwellers and even peasants. Some departments, even in rural areas, took the lead in this respect. Notable among them was the department of the Ain, where, from the beginning of the twentieth century onwards, water supply projects for both animals and human beings were extended to drinking troughs, wash-houses and street fountains. Nevertheless, towns and cities, particularly the larger ones, benefited from a sort of *de facto* bonus, in that their expenditure per head of population was necessarily lower as a result of the economies of scale that were possible.

The year 1820 marks the first real turning point in the development of urban water supplies. That said, however, the results actually achieved were modest; in 1892, less than half of French towns and cities distributed water under pressure, and only 127,318 of their total population of 4.5 million were consumers. In this respect, the name of Belgrand stands out. In 1847, with the construction of the Avallon supply pipe, Belgrand foreshadowed the building of the major aqueducts which were to supply the capital. Shortly afterwards, at the behest of Haussmann, he improved the quality of the water, moved the sewer outlets to a location downstream of the urban areas and installed separate networks of pipes for private and industrial use. He also introduced into France the mains drainage system pioneered in London, Brussels and Berlin. For the moment, the fear of disease was the dawn of wisdom. In Marseilles, large Roman-style aqueducts were constructed from 1849 onwards. In Angers, the old problem of the risks involved in drinking water from the Loire was raised once again. In Rennes, the famous fire of 1720 is said to have destroyed a fairly extensive water supply network. However, after many false starts, it was not until the 1880s that the capital of Brittany had an adequate water supply system.

The psychological factors that Guy Thuillier has pointed out were of considerable importance in this respect: until the end of the nineteenth century, water was far from being unanimously accepted. Nevertheless, the spirit of imitation – Paris copying London, Moulins copying Paris, Nevers copying Moulins, Clamecy copying Nevers – led to some very positive accomplishments. The false moves and gross blunders of engineers merely delayed matters; in Nevers, an engineer made a mistake in calculating the yield of a spring, and expensive and hastily constructed aqueducts were constructed to transport the water. The water catchments in Nevers, which were located downstream of the city's gelatine factories, floating wash-houses and sewers, inspire several colourful passages in Goubert's book.

As far as differences between and within departments are concerned, a figure relating to the Nord department should be noted in passing; in 1907, only 6 per cent of parishes had water, and of these 75 per cent were in urban areas. And yet the population of the department was not scattered; the large adjoining villages of the area would have merited the installation of a water supply even at that period. In this respect, the department of Aisne (to say nothing of Ain, which was still in the vanguard of progress) was significantly in advance of its more northerly neighbour.

However, we should not be deceived by the progress made in certain districts; in 1946, 58 per cent of the inhabitants of rural parishes were still fetching their water from their courtyard or from the village pump or fountain (the corresponding figure for town-dwellers of the period was only 4 per cent). It is true that market towns in the department of Lozère installed water supplies during the 1930, but in many villages the only places to have running water were hotels and other such establishments, where a proper water supply had become a necessity. All things considered, the change of attitude was gradual; the utopian dreams of the *ancien régime*, to have water in every home, were slowly being realized. All this had little to do with the many spectacular political and constitutional changes that had taken place in the mainstream of history since 1800.

A chapter on the socialization of the body and the items associated with that process gives Jean-Pierre Goubert an opportunity to recall earthenware washbasins, water jugs and even the brass fountains of the age of Louis XIII which, in the backward rural areas of eighteenth-century France, were the prerogative of a privileged few. During the Enlightenment, bathrooms were the de facto monopoly of the King of France and of certain aristocrats and wealthy members of the middle classes, who sometimes built their own private aqueduct to supply their bathroom! At a later date, the processes of hygienization and sanitization were to gather pace, starting with the families of future doctors and filtering down to those of future secondary-school teachers and finally to those of primary-school teachers.

Literature provides a few markers that should, of course, be used with caution. In Eugène Sue's *Wandering Jew*, for example, mention is made of a washstand, a zinc washbasin and a tap from which flowed an unlimited supply of water. Ethnology provides further evidence; in Minot in Burgundy, people washed themselves all over on Saturdays. Le Morvan, on the other hand, retained the tradition of not washing at all, except for the face, even as late as the 1880s. Even in Minot, however, optimism was not contagious; baths were just as likely to be used for storing potatoes as for bathing! The great spring and autumn washes provided an opportunity for the squire or lord of the manor to take advantage of the darkness of the wash-house to pinch the young washerwomen. Houses had no running water, and the chore of fetching water and wood was part of daily life in the French countryside.

Natural conditions should not be disregarded. If the eastern regions of France began relatively early to consume large quantities of water, this was in part at least due to the ease with which water supplies could be obtained. In Northern France, however, deep drilling was difficult. Polluted wells, on the other hand, were a persistent scourge which is reflected in various anecdotes. One of these concerns a well polluted by dunghills belonging to a neighbouring farmer; in another, a farmer complains that his tenants have failed to maintain a water tank in good repair. The prefect refuses to intervene in this latter affair; he hides behind the law, saying that it is only the cleanliness of animals that is at stake. Nevertheless, the spread of valuable knowledge continued; in the long term, it was irresistible. In 1899, following a typhoid epidemic, some villagers decided to take water from a spring that was purer than the water they had been consuming until then. In the department of the Rhône, the peasant tradition of cooperation was reflected in the communal effort put into the construction of a water supply system. On a more general level, the fear of drought and fire gave rise to serious consideration of the water supply problem. Direct contacts were made between the mayors of villages, hygienists and local doctors. The doctors supplied municipal officials with the necessary information and encouraged them to take direct responsibility for the management of the water supply. Similar developments were to take place from 1925 onwards among the 'interparish' syndicates.

With good reason, the author draws inspiration from the fine monographs of Le Play, whose *Les Ouvriers des deux mondes* is brimming over with information; the great sociologist has lost nothing of his topicality, despite his conservatism. He describes a sharecropper in the Confolentais region whose dwelling did not have the freshness of Flemish farms. The strange rituals intended to appease the saints of a paganizing Christianity are described by Le Play. They involved the use of flasks of holy water to wash the affected parts of the human body.

The reactionary thinker also describes, as a contrast to the 'primitive' sharecropper, the frequent ablutions of winegrowers in Alsace, although it is true that these mainly involved the head and the hands. In the large eastern province, hygiene was associated not with secular enlightenment but rather with devotion to the Catholic faith and to thrift, work and propriety.

The culture of the mid-nineteenth century defined the insalubriousness of towns and cities by contrast with the healthiness of life in the country; in many instances this was an illusion: village-dwellers certainly did not wash a great deal. However, the notion was not entirely false, since it may well have been the case that epidemics spread more easily in densely populated slums than in sparsely populated rural areas, where cottages were fortunately isolated from each other.

In any case, the countryside could not be treated as one single entity. A distinction must be made between the peasant's hovel described by so many authors and the parish priest's residence, which was the equivalent of the well-to-do farmer's country house; the priest's residence generally had a well and two lavatories. In towns and cities, things could be even worse; in a description of the Avenue de Choisy in Paris in 1882 there is mention of 'a house with a mud floor covered in tarred felt and in poor repair, the windows of which have no glass and are blocked up with scraps of muslin. The tenant sleeps on a bed of wood shavings strewn on the floor and pays the rent in advance. This house has no chimney, no water and no privy.' And in the rue du Château-des-Rentiers, pigs were being kept in bedrooms. . . . Although there had been a legal requirement since 1883 for all furnished lodgings to have a flushing lavatory, this was far from universally implemented. Shower baths, allotments, public wash-houses (and even, much later, swimming pools) spread gradually; they did not banish all the old hardships.

Nevertheless, the rural exodus was tantamount to an initiation into the civilization of water. The surveys conducted by Le Play once again provide our author with information on developments in the towns and cities; a socialist carpenter, recently settled in Paris, was dedicated to cleanliness, the work ethic and filial piety. A former seminarian, now married and working as a printer, benefited physically from a sense of moderation which enabled him to avoid serious illness; he made his children take a bath once a week in summer and once a month in winter; even more revealing is the disclosure that he had savings.

On a more collective level, capitalism was not always barbaric, or else it was fervently requested to cease being so; Republican legislation ignored the protest of landlords, who did not wish to be forced to connect their properties to the mains drainage system which was installed in Paris between 1865 and 1899. Nevertheless, for want of anything better, the chamber pot remained the alpha and omega of hygiene; until 1914, the fear of microbes was reflected particularly in the elimination of dust, which was believed to be the cause of tuberculosis. Thus one ancient myth was abandoned only for another to be created.

The birth rate also had a part to play. According to a survey carried out among shoemakers in the middle of the nineteenth century, it would seem that cleanliness declined as the size of the family increased and the parents had to share a small room with several children. However, the situation was not without remedy and was susceptible of improvement. In the case of printers in 1922, those conducting the survey concluded that sanitary standards were being maintained. This adherence to standards of hygiene was particularly strong in areas which already had high standards of hygiene and sanitation, such as those which had belonged for a time to Germany (as witness the case of the printers in Alsace).

In 1909, the factory inspectorate somewhat bizarrely charged a factory owner with sixty-five offences, one for each of his employees, because the toilets on his premises were poorly maintained. Was this excessive zeal justified in the face of a deplorable lack of hygiene? The lack of hygiene prevalent among workers of the period can be imagined from the unblushing description of toilets contained in a text quoted by Goubert. Sometimes there was even a total lack of toilet facilties; temporary workers in the brick works of the north and the sawmills of the southwest were sometimes housed in worse conditions than the horses, and there is no doubt that, under these circumstances, the forest served as a makeshift toilet. As a consequence, employers felt they were under no obligation to install the sanitary facilities required by their workers.

The situation was different among the higher social classes. Once again, Jean-Pierre Goubert makes judicious use of literary documents, including Michelet's *Journal*. Athénaïs, the wife of the great historian, practised hygiene in ways not so very different from those prevalent in the twentieth century. The toilet of Jules Romain's heroes in 1914 was as modern as possible, even down to the inclusion of a shower collar, which was praiseworthy in view of the limited financial resources of the young characters in the novel. Even Provençal literature has not escaped our author's attention, and he has used pieces by Victor Gelu, the writer from Marseilles, to describe the female companionship of the watering party. In Provence, the so-called hygienic bucket and, at night, an earthenware chamber pot completed the sacrosanct lavatory hut, which was usually situated at the bottom of the garden and closed with a door with a diamond-shaped hole cut in it.

The book proposes the term 'education of the body' to describe the process of hygienization and then goes on to examine a survey that was carefully conducted among educationists and doctors. This suggests that, in the final analysis, education actually played a relatively unimportant role in the diffusion of hygiene. On the other hand, the family certainly fulfilled some very important functions in this respect, with mothers playing a particularly active role. Historians and researchers in social sciences have for too long identified women with traditional cultures; in reality, they were instrumental, from the end of the nineteenth century onwards, in inculcating the correct attitudes towards hygiene among the well-to-do and educated classes. The final educational and medical survey examined by Goubert stresses the regularity of bodily hygiene. The stereotypical attitudes towards proper diet, adequate sleep, the benefits of physical education and the battle against fleas, alcoholism and even tobacco already formed an established part of the mentality of primary school teachers and doctors now in their eighties and nineties. Schoolteachers read specific texts to their audiences and taught their pupils a hundred ways of maintaining cleanliness and health. The inspection of hands and nails, the use of pointed matches, the carrying of toilet requisites and polishing of shoes are recalled with affection by the elderly correspondents questioned by Jean-Pierre Goubert. In sum, as the relationship to water changed, the policy of pious hopes that prevailed between 1750 and 1880 gave way to the construction between 1880 and 1940 of a vast socio-sanitary domain.

Jean-Pierre Goubert's research is entirely interdisciplinary, and his procedure is highly original. The first in his field, the author has at all points built up a study which never departs from its faithfulness to texts, documents and facts.

Emmanuel Le Roy Ladurie

Life
a little water
a few words on the tongue

Bernard Noël
(French writer born in 1930)

General Introduction
Water: quest and conquest

There can be few people in the developed world today whose homes are not connected to a water supply network. Indeed, we have become so accustomed to the presence of water in our daily life that it has been a long time since we have questioned its existence. Its social and hygienic qualities are recognized by most people, although not everybody washes all over every day,[1] and the rhythmic noise of the washing machine or the urgent gushing of a flushing toilet are now familiar to us. Our sense of modesty has become accustomed to this evolution in sanitary equipment; we are no longer alarmed by the nakedness of a body in a bath or under a shower. In towns, weekly cleaning of the gutters is no longer considered a luxury but a necessity.

This abundant supply of pure water that we use at all times throughout the day may well lead us to raise our habits to the status of norms and to make us forget how recent the conquest of water really is, and to what extent it ran counter to the sensibilities of our recent forebears in the last century. For a very long time, certainly for thousands of years, this behaviour, these habits, this relationship to water, did not exist. Water was a dominant force, imbued with powers, virtually immanent and generally feared. It was a symbol of severance and transition, particularly towards the beyond.

It was not until the nineteenth century that the old relationship to water began to be overturned and that this conquest, until then out of reach, was begun; the process was to continue until the eve of the Second World War. To judge from the work of some authors, this conquest seems to have inspired more than one vocation:

We were going to be schoolteachers and have the respect of an entire village and a life of luxury in a flat with running water. Everybody has choices to a certain extent: what tempted me was the running water, after ten years of fetching water in a bucket from a tank, of soaking wet clothes, of lost poles and, in winter, ice that had to be broken.[2]

To begin with, this conquest was a quest, and there is no better character than little Cosette in *Les Misérables* to symbolize this daily, ancestral act. Cosette, a servant and general dogsbody with the Thénardier family, fetches water in a bucket after night has fallen 'from a small spring halfway up a hill, near the road to Chelles, about a quarter of an hour's walk from Montfermeil'. The scene takes place in the evening, when the 'fellow' to whom aristocrats and bourgeois alike pay a farthing for a bucket of water, has already shut up shop:

She walked bent forward like an old woman, with the weight of the bucket dragging on her thin arms and the metal handle biting into her small chilled hands, pausing frequently to rest; and each time she put down the bucket a little of the water slopped on to her bare legs.[3]

The search for water was real drudgery. It was a task that sometimes fell to (unpaid) women and children and that was sometimes entrusted, for a small renumeration, to a public body. At the end of the eighteenth century water carriers were considered to be 'vile men and raucous women who upset the inhabitants of the districts in which these fountains are situated'.[4] There were almost 2,000 in Paris and about 20 in Chartres; they were often leaders of popular feeling and were at the head of the procession that marched on Versailles in October 1789. In 1843, one water carrier, Noël Taphanel, was one of the subscribers to the *Union ouvrière*, that was being republished by Flora Tristan. Loudmouthed and quick to use their fists, they were considered the dregs of the working classes. In *La Paysanne pervertie*, Restif de la Bretonne gives us a good idea of this in his description of Ursula who was forced into the humiliation of marrying a water carrier!

The water carrier treated me as his wife, or as his servant, he made me make his soup, I was forced to do his washing up, clean his boots and make his pallet [. . .]. A week went by without me being allowed to do anything other than dress myself in the most miserable, filthy, vermin-ridden rags that the poorest person would not pick up in the street and serve 'his grace', the water carrier.[5]

Until 1789, and for some time afterwards, the services of the water carrier were the most common means of obtaining water in the home, since there were very few buildings — either public or private — which possessed the water 'lines' granted by royal privilege in the case of France or by private companies in the case of London.

In Britain, development after 1760 was gradual. In France, the Revolution led to an abrupt change in this state of affairs. Access to water was a 'feudal right' which had been suppressed along with many others. Until then, it had been a privilege reserved for the nobility which brought them considerable revenue as a result of its economic uses in mills, dye works and fisheries. The main point is that, whatever the country in question, water became a good accessible to all.

Water, which had once been a gift of God or of nature and a privilege reserved for the nobility, now became the property of everyone and subsequently went on to acquire the status of industrial product. Nevertheless, this development took place extremely slowly between 1880 and 1940, and certainly less rapidly than the later expansion of the electric grid.

In reality, water continued for a long time to be considered as an essentially free gift from heaven, even if water carriers continued throughout the nineteenth century to sell it to the upper classes in towns. It was to require a whole programme of social and health education, pursued with increased vigour after 1880 by the press and the state schools, to persuade the working and even the middle classes that there were good reasons for using and thus paying for an abundant supply of clean water. For a long time, virtually everybody continued to be opposed to the frequent use of water; the reasons for this were numerous and included the fear of catching a chill, the indecency of personal hygiene, the absence of any perceived relationship between cleanliness and health, the poor quality and small quantity of water available (particularly in the countryside) and the custom of associating the use of the bath with a rite of passage (birth, marriage, death), or of basing the rhythm of traditional washing on the calendar of the seasons, with renewal following Jack Frost, the sap rising after a period of barrenness, life following death and hence cleanliness following dirtiness.

The slowness of this conquest can also be explained in another way; it required a considerable level of investment, even for a period of economic expansion. The capital sums required were all the more considerable because natural conditions were bad, the scattered population increased the per capita cost of connecting each individual dwelling, state subsidies were a long time arriving and because the machines, filters and reservoirs that were available for a long time produced variable amounts of low-quality water, depending on the season, which was not much appreciated.

This was why 'clean water was for a long time a privilege reserved for a few people, and until the end of the nineteenth century ill health was caused to a large extent by the poor water supply, its high price and a lack of hygiene, which contributed to the spread of epidemics.'[6] By charging for water while at the same time making handsome profits, the public, but more especially private companies were contributing to the slow democratization of water and making it easier for people to have access to it for their own use. Between 1800 and 1850, the price per cubic metre fell considerably for the urban consumer. The intellectual and financial elites thus joined forces in order to promote, secularize, medicalize and diffuse water. Everybody, technicians, architects, bankers, industrialists, hygienists, doctors and teachers, buried their differences and jealousies to participate in the conquest of water. They were its promoters, recruiting agents and builders.

This conquest had its epic moments and was characterized by the enthusiasm peculiar to neophytes. Thus some civil engineers claimed that the cast-iron conduits would last for a thousand years. It also had its conquerors, landscapes, monuments, incidents, aspirations and, of course, its scandals. There was the Nevers case, when it was realized that some subscribers were using three to five times more water than they were declaring; or the Paris case, when it was realized that the inadvertent mixing of clean and polluted water had led in about 1900 to a renewed outbreak of typhoid fever, and, finally, the London case when, during the second half of the nineteenth century, the famous *Reports* alarmed the city's inhabitants.

Over a period of two centuries, the conquest of water was reflected in feats and deeds, in acts and sensibilities. 'Since the family of gases and the race of acids and salts have been discovered [. . .], much thought has been given to the announcements of the chemists [. . .]. Water has begun to be analysed and we now reflect when we drink a glass of it, which our unconcerned ancestors did not do.'[7] Scientists began to analyse water from the end of the eighteenth century onwards and it became a subject of interest to monarchs, aristocrats and the middle classes; this last class often had handsome fountains in their homes which filtered and purified their water. Some anglophile French people expressed their desire for a plentiful supply of water. Thus in 1823 Achille Dufaud wrote to his father from Britain:

We lack one capital thing, and that is a water supply in the home. This provides great comfort, greatly facilitates the running of the household and makes possible a marvellous standard of cleanliness; with a compression pump and some lead pipes you will be able to make good this lack, but this expense is necessary for comfortable living.[8]

Engineers, hygienists, architects and administrators were all measuring the gap that separated the British tradition of hygiene, born of Puritanism and economic advance, and its French counterpart. The typical London house, for example, emerged as a veritable model of cleanliness, comfort and hygiene:

The kitchen is provided with two taps, one for hot water and the other for cold water [. . .] Without having to worry about it, Londoners can have hot water for their bath at virtually any time; and their wife or servant always has water available for washing or other domestic uses. The water closet is perfectly installed [. . .] There is always enough water in the bowl, not only to form an hydraulic seal and to avoid any smell, but also to prevent any fouling up.[9]

Water was thus a subject for conquest. Having been besieged by science and technology, it gradually became, in the course of the nineteenth century, an industrial and commercial product. This time, however, because it cleanses, because it rids the body of its own waste, because it 'purifies', it conquered the men who had subjugated it. The conquest turned out to be a double-edged triumph.

The coolness of water quenches thirst, its motion cleanses, its vigour gives or restores health. In a vaginal douche, it is privy to the 'darkest secrets' and, so it was believed, acted as a means of birth control; lukewarm and perfumed, it offers a pleasure that used to be reserved for dandies; in a village fountain it takes part in the life of the community and directs the ballet of the women who meet around it; in washbowls and basins, in bidets and water closets it permeates the rites of cleanliness and hygiene; in modern laundry, it provides us every day with white clothes; in the sewers it cuts through the entrails of the city; in a water tower, it changes the skyline.

Water thus conquered man in a triumph linked to increasing industrialization and an economy that devoured water. Nevertheless, it has retained its ancient power. In the words of Gaston Bachelard, it is still a 'substance of life and death' which gives rise to all sorts of rites intended to tame it; it has kept its ancient 'virtues' of quenching thirst and soothing pain, of maintaining health by separating the pure from the impure and ensuring long life. Its birth confers on it the gift of resurgence, even of resurrection. When it has disappeared, it reappears; once conquered, it reconquers, since it symbolizes the cycle of life and death.

This double-edged conquest was thus, at its outset, part of the domestication of water by the new knowledge and new technologies of the nineteenth century. Then, after 1870, during the *belle époque*, the mass diffusion of water developed in such a profound and durable way, due largely to the press, the schools and the hospitals, that it was finally water that conquered us by transforming the world and becoming part of our daily life.

As an industrial product, in the same way as coal, it ensured the growth of municipal companies and boosted the profits of capitalist companies, both small and large. Because the battle for the installation of hydraulic equipment was finally won, it altered the landscape, both above and below ground, mainly in the towns and cities. It gradually penetrated our daily lives: home, work, health, bodily pleasures – all emerged renewed if not always revitalized.

As a result of the 'model from on high' and because of urbanization, it finished by taking the citadel and imposing its new rites of cleanliness and hygiene[10] to such an extent that it made the 'superstitions' and behaviour of earlier times appear strange. Abundant and clean – at least in the France of today – it has become an everyday product that it is normal, even natural, to have in one's home.

Invisibility is indeed the height of conquest. Let us think about it for just a moment: who could seriously imagine a water strike?

PART I

Water, Purity and Hygiene

Just like the body with which it maintains such a privileged relationship, water has known different ages.[1] In its cosmological age, when the cult of the sacred magical fountains was celebrated, the healer and the water diviner were one and the same; in its religious age, which coincided in the West with the dominance of Christianity, baptismal water alone was able to cleanse the body of sin and contact between the naked body and hedonistic water was forbidden. 'Your heart,' wrote François de Sales in 1606 to Jeanne de Chantal 'must imagine only the "dew of heaven" and not the "waters of the world" '.[2] In an equally prudish style, *The Ladies' Doctor*, in the mid-eighteenth century, stipulated that 'in order to be naked without being seen, bath water should be clouded with powdered almond paste, bran and, in general, with any flour or resin dissolved beforehand in spirits of wine and then cast into the bath.'[3] Finally, there emerged a third age, which had its roots in the sixteenth century but did not develop fully until the nineteenth and twentieth centuries, in which there was greater emphasis on finding scientific explanations for the world. Water then fell prey to scientists: geologists, chemists, hydrologists and physicians. It underwent a process of secularization, becoming the embodiment of the increased importance attached to cleanliness before being seized upon by the ascendant medical profession to symbolize the hygiene that was sacrosanct to the followers of Pasteur and their rivals. Throughout the entire nineteenth and into the present century, the conquest of water forged ahead. The mechanism of conquest was based on water's ancient purity, and on innumerable purification rites – both pagan and Christian – and was solidly supported by a social code reflected in the attention given to cleanliness and tapped by the hygienist movement; water slowly conquered the world by permeating society and insinuating itself into innermost recesses of the body that had hitherto remained concealed.[4].

In each of these three ages, water was the subject of a different creed, each of which attempted to tame the power of the coveted liquid. In the cosmological age, this need to harness the power of water was expressed in 'water divining' and symbolized by the water diviner's dowsing rod; in the religious age, it was expressed through Christianity and symbolized by the cross of the Saviour (water purified of sin), while in the age of science it was reflected in the diversification of professional tasks and of scientific and technical knowledge, and symbolized, for example, by the doctor's staff.

However, it should not be imagined that these three ages simply succeeded each other chronologically. They all share some of the characteristics of the three ways of conceptualizing water and existence which still coexist today and are often even interdependent. All three of them are there, behind us and within us, close enough to touch each

A descendant of the stoup, or holy water basin, at least in shape, the domestic fountain was in widespread use among the well-to-do classes in the eighteenth century, providing a link between the sacred and the profane uses of water.
(Reproduced by kind permission of the Musée National de Céramique, Sèvres. Photograph: Réunion des Musées Nationaux.)

other, brothers at war with each other in often relentless conflict whenever the third age, with its would-be rigorousness, pronounces judgement on the *superstitions* (a convenient word which does not encourage reflection) of the first or second age.

1

New knowledge

The water diviner and the engineer

It is very difficult to state exactly how far back in time the use of the divining rod goes, or who the first water diviner was, unless he is identified with God himself. The first authenticated document is a German engraving dating from the early fifteenth century which depicts a miner prospecting with a divining rod in his hands. As a symbol of power, including both sexual and political power, the rod or wand was common to the Pharaohs of Ancient Egypt, to Poseidon, the Greek god of the sea and springs, to Athena, who used it to age or rejuvenate Ulysses, to magicians (such as Circe) and the prophets of Israel such as Jacob, Aaron and Moses. Tacitus tells us that the Etruscans used the *litius*. And when the Roman legions marched into Gaul and Germany they were preceded by 'water diviners' whose task was to detect the underground water necessary to supply the troops. The diviner's rod was always made of wood, which both symbolized the relationship with the visible world (branches) and the underground world (roots) and distinguished it from the royal sceptre or the bishop's crosier. Indeed, it possessed the active power of *revealing* a hidden element: water, mines, treasures, the future. It was cut in accordance with precise rituals; for example, at sunrise, from a branch held in the left hand and then cut with three strokes of a pruning knife after the incantation of a ritual formula. The magic rites were condemned very early by the Church, including Luther in 1518. In the fifteenth and sixteenth centuries it was used to locate deposits of metal ores or to determine the site or extent of a seam, particularly in Germany.

In the seventeenth century, however, scholars and scientists attempted to explain the phenomenon of the divining rod rather than simply describing it. Let us examine the case of France. The Jesuits became interested in it and no longer condemned it as openly after having seen the rod turn in the hands of a 'pious and honest man'. However, this period was short lived, and it was not long before things began to deteriorate for water diviners. At the end of the century the Aymar affair occurred. Jacques Aymar, who was a rich peasant from the Lyons area, began to acquire a reputation when, in 1688, he claimed to have exposed a thief and even to have identified a murderer by use of a divining rod.[1] In the presence of the court, he underwent a whole series of experiments intended to verify his powers but which in fact went against him. This outcome was immediately seized upon by opponents of the diviner's rod and all those who did not accept that he could be used for 'moral purposes', who loudly proclaimed that the 'diviner was unmasked'. In 1689, concerned by the importance assumed by the famous rod, Father le Brun requested a consultation with Malebranche. The philosopher accepted that the rod turned in the presence of natural elements − subterranean water or minerals − but, on learning that it also turned in the presence of thiefs and murderers and 'being unable through thought alone to explain such effects, he attributed them to the Devil'. Malebranche's conclusions and those of Father Le Brun were generally accepted by the scientific elite, and the divining power of the rod was ascribed to the Devil.

The polemic reappeared when, several years later, the abbot of Vallemont published a book entitled *La Physique occulte* which contradicted the diabolic interpretation which had prevailed until then. Vallemont was a supporter of the corpuscular theory, according to which each particle of matter, including water and minerals, produced emanations which could be taken up by other materials, particularly the diviner's rod. The publication of his book brought immediate protests from opponents of the divining rod. Father Le Brun assumed the leadership of this crusade. In 1701, the 'Holy Inquisition' condemned Vallemont's book. Silence then reigned for almost eighty years.

It was not until 1781 that the diviners were brought back into the public eye as a result of the extraordinary gifts of a certain Bléton which were studied by Pierre Thouvenel, a doctor in Nancy, and subsequently verified by the French astronomer Lalande. Like his Spanish counterparts in the sixteenth century, Bléton worked *without* a rod. When he arrived in a divining zone his whole body began to shake; it was in this way that he detected the lie of the Arcueil aqueduct before a crowd that *Le Journal de Paris* of May 1782 put at 5000 people.

For Dr Thouvenel, a great admirer of Mesmer, there were 'obvious relationships' between animal electricity, magnetism and the Bléton 'phenomenon'. There then quickly followed a series of different interpretations, with each scientist being quick to establish a link between the scientific theories of their choice and the 'rational' explanation of the phenomenon of divining.

Between 1806 and 1826, Count Tristan, who was acquainted with Thouvenel's works, carried out new experiments with the divining rod. For his part, he believed that he had found the explanation for the rod's rotation in a theory involving electricity. According to him, the rod acted like the needle of a galvanometer. When the 'fluid' was positive it flowed into the diviner's right hand and the rod pointed upwards. When the 'fluid' was negative, it flowed into the left hand and the rod pointed downwards.

The Baron de Morogues continued his uncle's research. In 1854 he published his *Observations on the Movements of Divining Rods and Pendulums*, in which he developed the 'electrical theory'. He put forward the hypothesis that all existing bodies were surrounded by 'electrical spheres' which acted in particular on the sphere of the diviner himself whose muscular contractions were recorded by the rod.

The discoveries of Galvani and Ampère, who described magnetic and electrical fields, delighted all those concerned to provide a 'rational' explanation of the phenomenon of divining. The electromagnetic theory had its hour of glory and gained many followers, in particular Father Carrié, the parish priest of Barbaste, who in 1863 published his *Art of Discovering Springs by Electromagnetism*, and the quartermaster Jansé who, between 1874 and 1884, produced several methods for estimating the depth and yield of a spring.

Around the turn of the century, discoveries in the areas of radioactivity, radiation and radio waves led to the appearance of new theories. The press became interested in the phenomenon, the number of articles increased, and in 1913 the International Union of Water Diviners was founded.

Although the Great War brought studies of the divining rod to a halt, diviners were used to discover subterranean caves that might be used as shelters by the enemy. The aid of Father Bauby, the famous water diviner and inventor of the term 'radiesthesia', was enlisted in detecting shells buried in the ground.

Once the war was over, the investigations were resumed, but the nature of the explanation changed. It soon ceased to be physical and became psychic, notably as a result of the influence of Father Mermet. This enabled the scientific world to cloak divining rods and pendulums, radiesthesists and water diviners in a veil of obscurantism. This attitude still persists today, as shown by this letter from a present-day civil engineer in government service:

After thirty-two years of experience, it is my view that calling on the services of a 'water diviner' is analogous to painting the face or other parts of the body and dancing around to the beat of a drum in the hope of making it rain. I make no value judgement of this method, which may be effective when there's thunder in the air.[2]

Thus the dispute is still simmering today and the scientific explanation is undergoing critical scrutiny. However, the part played by magnetic fields was confirmed in 1962, when a nuclear magnetometer was used to reveal the existence of a very weak magnetic anomaly to which the water diviner reacts in a divining zone. The relationship between magnetic fields and 'divining signals' highlighted by the Frenchman Yves Rocard has also been confirmed by studies carried out by Russian and US scientists.[3]

Water and the sciences

In the nineteenth century, however, the water diviner's rod and the mysterious side of his 'gift' were in conflict with the explanations offered by scientific knowledge. The activity of scientists was devoted for the most part to penetrating the mechanisms of water. Their aim was not only to discover its cycle and natural system, but also to determine its physical, chemical and bacteriological composition.

The hygienist ideology which emerged in the Western countries between 1750 and 1830 thus gradually acquired a scientific base solid enough to enable its adherents to put forward technical proposals for the gradual extension of a clean water supply, as well as solutions to the equally fundamental problem of the disposal of waste water.

The hydrologic cycle

The current theory of the hydrologic cycle, that could be described as 'atmospheric', remained controversial until the middle of the eighteenth century. At that time it was opposed by the theory of the subterranean cycle. Based on Aristotle and Plato, sustained by the Judaeo-Christian tradition and supported by the story of the Flood, this theory stated that the sea, in which the overheated centre of the earth was steeped, was the source of all water. As a result of evaporation and capillary action, the water rose as far as the 'ceiling' of the caves which formed the substratum, thus forming the rivers that then flowed away towards the ocean. According to this theory, water was purified by the fire burning at the centre of the earth and protected from pollution by an impermeable layer which extended downwards for several metres. Water pollution was a phenomenon that could not exist.

In contrast, the theory of the atmospheric cycle, which prevailed after the brilliant demonstrations by Clairaut in 1743, and which was confirmed by Buffon's arguments, stated that all water 'even the purest, contains germs from a host of miscroscopic animals that develop over time, depending on variations in temperature, movements, the nature of the mud, etc.'[4] Because the earth's internal vault does not exist, any more than its hollow interior is filled with water, the hydrologic cycle can only be atmospheric: 'Rain penetrates the earth as far as the impermeable layer; there it stagnates or flows away to form springs and then rivers, and evaporated sea water produces clouds.'[5] For chemists and doctors, water became, like air, the polluted element *par excellence*, the favoured place for the 'spontaneous generation' of microbes, a theory supposedly confirmed by the serious cholera epidemics that broke out in the course of the nineteenth century. It was to take all the perspicacity of Koch and Pasteur to demonstrate the stupidity of this theory and that microbes did not breed in pure water. In the meantime, the danger was there, omnipresent, invisible, and it was necessary to fight for the health of human beings against the noxious, supposedly fatal, vapours given off by water.

The anlaysis of water

The deliberations of François Lavoisier on the *Nature of Water* (1770), in which he set forth the basic principle of modern chemistry – that of the conservation of energy – constitute, together with the chemical analysis of water, one of the outstanding discoveries of the late eighteenth century. In other words, the end of the eighteenth century heralded a break with the science of ancient and medieval times. Until then, scientists had rallied round the idea, which stemmed from Greek philosophy, that water was one of the four natural elements on which all the others were based. Macquer and Sigaud-Lafon anticipated the decomposition of water by discovering that, when hydrogen was burnt under bell jars, droplets of water formed on the walls. In 1781, Priestley conducted an experiment the results of which pointed in the same direction. In the same year, Cavendish repeated this experiment on a larger scale. He obtained several grams of the precious liquid and deduced from this that water was a combination of oxygen and hydrogen. While the British scientists were engaged in these experiments, the French scientist Monge was obtaining identical results and coming to the same conclusions. Several years later, the French scientists Lavoisier, Lapalce and Meunier conducted some new experiments. By burning large quantities of oxygen and hydrogen, they succeeded in showing that their total weight was equal to that of the water produced by the process of combustion. Many scientists then embarked on similar experiments of their own. The results were conclusive: H_2O was born! Even more importantly, the publication by Lavoisier in 1789 of his *Elementary Treatise on Chemistry* threw the science of his day into confusion, not only chemistry, but also agronomy, industry, hygiene and medicine. The principle of the conservation of energy, the introduction of quantitative measurement and, in particular, the notion of the chemical equation, found a powerful echo among Scandinavian and British scientists. The British scientists Richard Reece (1775–1831) and Samuel Parkes (1761–1825) became propagandists for the 'New Chemistry'. In 1808 Reece wrote: 'at the present time, a new era may be struck by the introduction of French chemistry into medicine.' Taken in conjunction with John Pringle's brilliantly intuitive idea about the dangers attendant upon animal putrefaction, Lavoisier and his rivals had given the decisive finishing touches to the understanding of the physico-chemical properties of water; it had now become possible to control, purify and preserve water and also to make some improvement in standards of health. Not content with understanding natural mechanisms, Lavoisier and the scientific profession attempted to give their discoveries a social value. In 1818, the Englishman Parkes wrote:

The human body is itself a laboratory, in which, by the various functions of secretion, absorption, etc, composition and decomposition are perpetually going on: how, therefore can [the medical student] be expert to understand the animal economy, if he be unacquainted with the effects which certain causes may chemically produce? Every inspiration we take, and every pulse that vibrates within us effects a *chemical* change upon the animal fluids.[6]

Thus the schemes proposed by all those interested in the future of water purifying machines landed in large numbers in the offices of local authorities and scientific associations. Sailors were particularly interested in better means of keeping water fresh on long voyages.

Various methods were tried out, including adding sulphur to the barrels (Deslandes), putting coal in the holds of vessels, as recommended by Berthollet, and the use of various salts, particularly manganese peroxide; each of these methods had their merits; today [1885], the preferred method is to use metal tanks, which were devised in England and which have been statutory in the French merchant navy since 1825. . . . In addition, the French navy has begun to use distilling machines which prevent men from dying of thirst at sea. The machine in current use is the Perroy device which aerates and filters at the same time as it distils; it is capable of supplying 20,000 litres of water in twenty-four hours; the cost price is 0.015 francs per litre.[7]

In the eighteenth century, distilling machines had not been perfected to that level. In Britain, there was concern about the quality of water and its effects on health. The tastes of Londoners living in the smart residential districts were becoming more refined, and a private company built reservoirs in order to purify the water that it distributed. Charles Lucas, an exiled Irish patriot who had qualified as a doctor in Paris, wrote in 1746: 'Water, the most useful and necessary part of creation, whether economically, physically or medicinally considered, has been so far and so long neglected.'[8] Making use of his five senses and the little prescientific knowledge that he had, he concluded, not without satisfaction, that, due to the abundance and variety of its water, 'our capital is the most healthful great city of the world.'[9]

There has not been a period when the need to ensure the purity of water has not been recognized. The senses common to both man and animals (particularly taste, sight and smell) have generally been used to reveal both the merits and the deficiencies of the coveted water. These criteria, that some label 'primitive', are not as out-of-date as is sometimes claimed today. In any case, they were certainly of great use to our ancestors in the West who worshipped tradition and who ensured that knowledge of the 'nature of water' was handed down orally from generation to generation. At the end of the eighteenth century, the vegetable boiling test dear to our grandmothers was considered an excellent test. In 1798, Dr Joseph Browne of New York wrote: 'for water that contains any of the fanets with an earthy base, such as nitrate of lime and magnesia, will not do well for either of the above purposes.' In short, he maintained that water 'that is clear and from a running source, that boils leguminous vegetables tender, in which soap readily dissolves and has no bad flavour, may be pronounced good water'.[10]

Towards the end of the eighteenth century, the consular assembly in Aurillac, a small town in the heart of the Auvergne noted in its proceedings that:

It would be necessary to have it [the water from the fountains and springs] examined by more expert doctors in order to decide whether it is wholesome and free from contamination and incapable of spreading disease among those inhabitants who might drink it.[11].

In his description of Rennes in 1789, Dr Salmon considered the city's water to be 'hard, sulphorous and unsuited to use in kitchens or for cooking vegetables'. Not without justification, he attributed these shortcomings to the fact that the water 'is filtered through ground containing slate, a sort of flaky rock known to naturalists as schist and a lot of tuff'.[12]

In fact, the criteria for assessing the quality of water were not soundly based. Even scientists were reduced to appealing to experience and the senses, in a word to that empiricism that they so often spurned. Nevertheless, many of them, notably doctors or 'enlightened men', deplored the fact that their advice on the use of water was not heeded. Against that advice, water was frequently drunk straight from wells, even in towns, and only the more well-to-do could afford the supposedly better water purchased from water carriers. This grievance is reflected particularly clearly in the writing of Dr Moulenq. On the subject of Valence, he wrote:

Despite the ease with which water can be obtained from the fountains that serve the various districts of the town, the inhabitants find it more convenient to use water drawn from wells, of which they make greater use than that taken from fountains.[13].

The French doctor Mignot Degenets turned this grievance into condemnation when he set the stamp of shame on an 'old habit' of the nuns of Thiers who were the only people to use water from a well:

Thus congestion, obstructions of the viscera and the lower abdomen, swelling of the hypochondrium and all the cachetic diseases abound there, and many young nuns perish in the first five or six years after taking the vows.[14]

However, the quality of water analysis varied a great deal from one doctor to another. The two doctors quoted above were merely run-of-the-mill. A minority, however, better informed about the latest developments in chemistry, did stand out from the rest of their colleagues, particularly the Rouen doctor Lepecq de la Clôture. The criteria that he used were much more subtle and expressed in a quantified way. Thus the solid residue was recorded in grains per pound and per pint, as were the substances contained in the water: chalky soil, calcium sulphate, sodium chloride and potassium nitrate. In his summary of his own results and those obtained by Décroizilles, Lepecq drew up a classification of the various supplies of water available in Rouen in 1778. He wrote: 'The water taken from the Seine at the Rouen bridge is much purer, less contaminated by foreign substances and lighter than the water in our springs and fountains.'[15] And, in order to pronounce a judgement on the Rouen water supply, he compared the weight of the residue with those calculated for Paris by the commissioners of the Academy of Science. He did not, however, scorn the evidence offered by experience, since he also lists the dissolving of soap in water as a criterion of quality. Thus, confident in his own knowledge, Lepecq concluded, as a health expert:

It would thus be a very good idea not to use the Notre-Dame spring for drinking, cooking or as a medicament. And we think it best to advise the inhabitants of the Saint-Nicaise district to walk a little further and draw their drinking water from the Fontaine de la Croix-de-Pierre and the one in the rue de l'Epée and to use water from the du Plat spring only for cleaning and washing clothes.

This distinction between water for drinking and cooking, and water for other purposes, seems to us today to be an obvious one. In fact it dates back to the end of the *ancien régime* and, in French towns, had its origins in the old system of distributing water by water carrier, whereas British cities, such as London, had had a system for distributing non-purified water by pipes since the 1560s.

A similar distinction between pure and waste water became widespread in the eighteenth century. In 1762, P. Patte, in a project drawn up for the city of Paris, proposed that the water should be fed into two channels running parallel to the Seine in such a way as to separate water to be used from waste water. Projects of this kind proliferated in the second half of the eighteenth century. They were intended to protect towns from water shortages and to rationalize the distribution system by studying the substrata. In 1742, L. Buache, in a personal submission to the Academy of Sciences, used his observations on flooding in the Seine valley to draw conclusions on the nature of the substrata that were used until the Second Empire. There were also plans to increase the volume of water distributed in the cities, to improve the distribution system, particularly in Paris, and to make provision for the disposal of the filth produced by 600,000 inhabitants, to say nothing of the horses! However, there was not yet any question of connecting individual households to a mains supply or installing British-style toilets. This whole programme was drawn up within the Academy of Sciences. After 1742, the King entrusted this task to the Academy, thus divesting the city authorities, architects and the police of their responsibilities in this area. However, the Academy of Sciences was not the only expert body to be consulted, since the provincial academies were also called upon. The academy in Lyons was particularly active in this respect. Water was even chosen as a subject in academic competitions: between 1770 and 1775, there were nine studies relating to means of distributing water in towns, and between 1778 and 1780 there were four relating to 'the crossing and cleaning of the city'. At that time it was recognized that 'the city has to be cleaner' and the Academy 'shares the aspirations of classic urbanism towards salubriousness and wishes to provide the debate on architecture with a technical foundation'.[16] In this respect, the competitions held by the Academy occupied the vacant ground that was to be filled in the nineteenth century by professionals, administrators and politicians. As a result of the Academy's competitions, the number of proposals put forward began to proliferate from the end of the eighteenth century onwards. However, only very few were ever realized, because of a lack of finance and, more particularly, an absence of political will.

Indeed, although the Governments in power between 1780 and 1810 were convinced that 'contaminated water is one of the commonest causes of sterility in women, of endemic diseases, of autumnal fevers and of epizootic diseases' and that 'in accordance with the opinion of Hippocrates, it is one of the principal causes of the depopulation experienced in certain regions,'[17] they did nothing to develop charcoal filters, except in the case of water taken on board warships, first in the British navy, then in the French. With few exceptions, neither central

Government nor local authorities played any part in laying down standards to control the quality of water.

In the early nineteenth century, the state did not bother to intervene in the area of prevention. The welfare state was still a long way off, as is shown by the reply given by the Consultative Committee on Arts and Manufactures on 16 March 1819 to 'Mr Rambaux, retired officer':

The author would wish to have water purifying stations constructed in many parishes in the kingdom, like those in Paris, and he even wants this purification to be carried out on a large scale in order to supply water not only for people but also for animals, since he claims that bad water is inimical to health. . . . His plan, which he discusses at great length, does not merit Your Excellency's attention. It would doubtless be a fine thing if there were to be more water purifying stations; however, the government will play no part in this: it is up to private individuals to establish them; the government can only protect them.[18]

Between the end of the eighteenth and the beginning of the twentieth centuries, science and technology evolved considerably. This evolution is reflected in the development of techniques for the analysis of water. After 1880, chemical analysis, which is undeniably very useful, was recognized as insufficient in itself. It was generally considered that it had to be complemented by bacteriological analysis, in accordance with the requirements laid down by the scientific organizations. The guiding principle of such analysis was that 'drinking water should not contain any mineral, organic or inorganic substance that may harm the organism that absorbs it?.'[19] At the beginning of the twentieth century, these mineral substances were lead, copper and arsenic, while the harmful organic substances were microbes, microorganisms and certain vegetable substances.

The theory that harmful substances were present in ground water had become virtually consistent with contemporary theories. Man and animals deposit waste on the surface of the earth which acts as a vehicle for germs and parasites. The rain water that seeps through the soil to be held underground takes with it these pathogenic and dangerous microbes. This contamination can sometimes be avoided since some soils have the ability to act as filters and thus to prevent the microbes from reaching the ground water. It is then simply a question of testing its effectiveness before pumping up the ground water.

However, ground water was not the only source of water in use at the beginning of the twentieth century, although it was preferred by scientists. Surface water from rivers and lakes was used as well, and this also of course contained many impurities. Similarly, the organic matter in such water is filtrated naturally by a fairly active process of nitrification that can be revealed by chemical and bacteriological analysis. This nitrification was not, however, considered sufficient to

'Mastery over water'. Mobile laboratory for the analysis of water. Tracy-le-Mont (Oise), 1919.
(Photograph: © Roger-Viollet.)

enable surface water to be used as drinking water; and, in contrast to the previous period (sixteenth to eighteenth centuries), preliminary treatments were judged to be necessary, usually filtration through sand. These procedures, which originated in Western Europe, spread to the United States in 1869, when James P. Kirkwood published his *Report on the Filtration of River Waters, for the Supply of Cities, as Practised in Europe.* In the same year the state of Massachussets established a State Board of Health and in 1878 sent an eminent scientist to Europe to study the techniques used to purify water. Five years later an experimental water purifying station was built. As a result, the link of causality was established between a pure water supply and standards of health. Water supply systems were then built in towns and cities throughout the United States even more rapidly than in Western Europe.

Water in crisis

After 1800, the quality of water in the major industrialized countries had, in general, deteriorated seriously. The medical committee set up to analyse the water supplied by the public pump in the Impasse de la Porte Bleue in Brussels wrote: 'it is lemon yellow in colour, with a very faint marshy odour and an insipid taste; it is clear and does not become cloudy until it has stood for three or four days, when it leaves a slight whitish-grey deposit on the sides of the container. When boiled, it gives off an unpleasant smell and the surface becomes covered with a pearly film slightly tinged with blue.'[20] Even worse was the water drawn from the 8,027 wells that formed the basis of the Brussels water supply system between 1830 and 1850. It had a 'disgusting flavour', a 'foul odour', an 'extremely disagreeable smell of rotten wood' and a 'nauseating taste'.[21] In Paris, an expert of the period (1844) concluded that barely 10 per cent of the water drawn from fountains in the city was drinkable![22] In London, the quality of the water was even worse, despite the fact that the city's water supply network was one of the first to be installed, due to the establishment of private water companies. Londoners complained of finding 'leeches' in their water, as well as 'small jumping animals that looked like shrimps', 'an oily cream', a 'fetid black deposit', etc. In 1827, the question was brought into the public arena following revelations made by a director of one of the water companies. Meetings and petitions were organized and a Royal Commission of Inquiry was set up in 1828. In the same year, the Commission reported that the Grand Junction Company was taking water from the Thames at a point opposite one of the largest sewer outlets in London, that the reservoir of the New River Company contained 8 feet of mud and that it had not been cleaned for 100 years, and that the Middleton aqueduct, which had been neglected for 200 years, had become a ditch into which drained waste water from the villages on its route. However, it was not until Edwin Chadwick's famous Sanitary Report was published in 1842 that the problems of supplying water to urban districts were fully recognized.[23]

In New York, the need for a public water system had been recognized since before the War of Independence. However, on the eve of the nineteenth century, no progress at all had been made, although many proposals had been submitted for consideration. Some New Yorkers began to lose patience with the delay and expressed their bitterness at paying high prices for the impure water they received from the Tea Water Pump! In the *Commercial Advertiser* of 5 September 1798, one correspondent ridiculed the complacency of those who pretended that the city's water was pure.[26]

In a similar vein, the poet Coleridge in 1828 painted a sad picture of the 'romantic Rhine':

> In Köhln, a town of monks and bones
> I counted two and seventy stenches,
> All well defined, and several stinks!
> Ye Nymphs that reign o'er sewers and sinks,
> The river Rhine, it is well known,

Doth wash your city of Cologne;
But tell me, Nymphs, what power divine
Shall henceforth wash the River Rhine?

In the German towns of Westphalia, which experienced a new wave of industrialization after 1850, the situation gradually deteriorated. In Bielefeld,[25] the population increased rapidly (5,581 inhabitants in 1798; 10,706 in 1850; 18,693 in 1870; 63,007 in 1900) and the expanding new industries devoured ever greater quantities of water. The pollution caused by the growing population and, more particularly, by the expanding industries, forced the city authorities to change their policy on water supply and purification and to consider spending considerable sums on improvements. In 1875, a medical report condemned 'the water in the Lutter stream [which is] a milky or brownish colour [. . .] because of the effluent discharged into it from various factories, particularly the laundries and linen factories.'

In Cologne, the poor quality of the water from the Rhine was showing the effects of the rapid growth experienced by the city. In less than a century, the population had increased almost fivefold:

Year	No of inhabitants
1834	60,000
1861	120,000
1880	145,000
1888	176,000
1890	266,000
1910	277,000

Industry was expanding, diversifying and modernizing: sugar refineries, factories producing machinery, engines, wagons and cables, paper, dye and textile factories and a large number of tanneries were all more or less polluting and devoured vast quantities of pure water.[26] Examples of this kind could be found in all the major industrial regions of Western Europe and the east coast of the United States.

A remedy: the hydrometer

In 1856, the Frenchman Bontron developed a hydrometer which made it possible to determine the hardness of water. By this means, the amount of lime and magnesium sulphates was measured in relation to a soapy solution. The acceptable upper limit of total hardness was around thirty degrees. This method, proposed by Bontron and Bondet, arose from observations made by the English doctor Clarke of the use of a solution of soap in alcohol to measure the hardness of water. This method 'is based on the well-known ability of soap to make pure water frothy and to produce froth in water containing salts, particularly those derived from lime and magnesium only to the extent that these salts have been broken down and neutralised by an equivalent proportion of soap, leaving a slight excess of soap in the liquid. Since the hardness of water is proportional to the salts that it contains, the quantity of soap solution required to produce the foam provides a measure of its hardness.'[27] With the exception of mountainous regions such as the Vosges, the Alps and the Highlands of Scotland, there were considerable variations in the degree of hardness of water. In Paris, it fluctuated between 2° and 128°, between 13.5° and 135° in Lyon, between 1.5° and 44° in Strasburg and between 11.25° and 48° in Rome. In contrast, the hardness of the water supply in some British towns was relatively low: between 11° and 16° in London, between 12° and 15° in Liverpool, 12° in Manchester, 5° in Edinburgh and only 1° in Aberdeen.

Of greater concern was the fact that the inhabitants of some districts preferred water with a degree of hardness in excess of 100°.[28] However, some chemists considered that water with a hardness of between 35° and 40° could make excellent drinking water. In 1864 the Frenchman Peligot, who was in charge of the water supply in Le Havre, described the town's water as 'abundant, fresh, clear, with an excellent taste'.[29] Nevertheless, the fear that serious diseases might develop still persisted; it was even prevalent in scientific circles when, in the mid-nineteenth century, the hardness of water was examined. Some British doctors were of the opinion that:

water containing carbonate of lime facilitates the production of obstructions of the viscera by reducing natural secretions; this results in a state of chronic constipation, which is obviously injurious to health. The Glasgow Medical Society, which is well placed to examine the comparative effects of hard and soft water, since the water supply in the northern part of the city is hard, filtered water from the Clyde (25° on Bontron's scale, 15° on the Hardness scale), whereas the water in southern part of the city is soft water from the mountains (5.6° on Bontron's scale and 4° on the Hardness scale), is unanimous in its recognition of the beneficial effect on public health of the introduction of this soft water for domestic purposes'.[30]

Water and disease: the examples of cholera and typhoid fever?

The 1900s saw the beginning of water's *belle époque*. Thanks to the development of quantitative analysis, physico-volumetric methods and germ counts, it was now possible to analyse water effectively, and it was now supplied to an ever-increasing number of customers in conditions of safety and convenience that continued to improve. As a result of the discoveries of microbiology, death and disease had finally been banished from water.

However, it was not until the beginning of the twentieth century that the aetiology of some infectious diseases was discovered. It was then realized that some germs, including the typhoid bacillus and the cholera bacterium, thrived in water, a discovery that sowed widespread terror.

Water and cholera Initially, at the end of the eighteenth and the beginning of the nineteenth centuries, the theory that 'miasmas' caused serious illnesses hastened the construction and even the completion of some water supply systems. This was the case in London, Paris, New York and Philadelphia. In the United States, 'water engineering schemes were implemented for cleaning the streets, in order to rid them of the rubbish which, it was thought, was the cause of the epidemics of yellow fever and cholera. In the absence of any scientific evidence for the relationship between water and disease, the demonstration effect was considerable and gave rise to new schemes. New York launched the Great Croton project after it became known that Philadelphia, which had a vast municipal water supply system, had had lower morbidity and mortality rates during the 1832 cholera epidemic. The connection between water and health was further highlighted by the work of John Snow, a London doctor who, in 1854, traced the cause of the cholera endemic in Soho to a single polluted well, the famous Broad Street pump.[31]

In 1852, just as the city of Brussels was seriously thinking of providing itself with a water supply, information was requested from the Brabant Medical Committee on 'the general standard of health of individuals using water from springs that surface in certain districts of Nivelle and Brussels'. The Medical Committee concluded that 'in any case, the influence of the water is real and generally harmful in the districts listed as having unhealthy water [. . .]. Moreover, is it not known that it is precisely where there is a lack of pure water that diarrhoea, dysentery, typhoid fever, intermittent fevers, etc are most prevalent and that it is there that certain epidemic diseases cause the most terrible losses of life.'[32]

Generally speaking, however, cholera did not come into the argument, either in Paris or London, Brussels or New England. The scientists stuck to their opinions. There was no evidence that poor quality water was the direct cause of the disease, even on the local level. Until 1883, few doctors, particularly in the United States, accepted the notion that the cause of cholera lay in a microscopic organism that had never been seen! And with good cause. . . . At the very most, the Health Committee set up in Paris in 1832 recommended the use of filtered water. As a consequence, several profiteers extolled 'the sale of purifying carbon filters (Ducommon, 6, boulevard Poissonnière), stating quite definitely that the Medical Association guaranteed that their use would ensure health and life'.[33] Indeed, 'the committee on cleanliness recommends that the streets should immediately be made cleaner, houses healthier and the existence of their inhabitants better.'[34] However, for lack of scientifically based evidence, the development of water engineering hardly kept pace with industrial and demographic expansion. 'In April 1854, when the cholera was spreading for the third time, the engineer Eugène Belgrand suggested that the sources of the Dhuis and the Marne should be diverted towards Paris to join up there with the filtered water of the Marne. The project was not taken up until 1859.'[35]

In Britain, despite the Sanitary Report of 1842 and the survey carried out by the General Board of Health in 1848, the Metropolitan Water Supply Act of 1852 did not follow the recommendations issued by the General Board of Health and the London water supply was not extended to Farnham or Surrey Sands. The three experts in chemistry who were consulted in 1851 suggested that this measure was unnecessary: 'The river may reasonably be supposed to possess, in its self-purifying power, the means of recovery from amount of contaminating injury equal to what is present exposed to its higher section.'[36] Chemists still believed that, as far as organic contamination was concerned, 'the indefinite dilution of such matters in the vast volume of the well-aerated stream is likely to lead to their destruction by oxydation, and to cause their disappearance.'[37] However, the private companies in London were asked to cover their reservoirs and to filter their water effectively before distributing it (Metropolitan Water Supply Act 1852). It was not until the end of the nineteenth century that the measures implemented brought about any significant improvement in the quality of the water supplied and that complete reliance ceased to be placed in the self-purifying capacity of rivers. . . . And in Munich, the effects of the measures implemented on the order of von Pettenkoffer were not felt until the end of the century. The mortality rate for typhoid fell from 0.72 per 10,000 in 1880 to 0.14 in 1898. In Cologne, the 1832 epidemic claimed only one victim. However, while discussions on the provision of a water supply were proceeding at a good pace, the 1849 epidemic caused 1,357 deaths. 'The year of the cholera', as it was subsequently to be known, claimed 45,315 victims throughout Prussia. Even though Cologne was virtually spared by the other epidemics (except in 1866), the threat of it continued to hang over the city.[38] A whole group of scientists, doctors and engineers began to think seriously about the causes of this 'disease that spread terror'. In 1865, while the threat of a new epidemic coming from France hung over the city, Dr Eduard Lent, a member of the Cologne Medical Association, set up a multidisciplinary committee. He called it the Public Health Committee and divided it into four subcommittees which analysed in great detail the physical, social and sanitary situation of the city. In March 1867, the Committee submitted a report to the municipal authorities in which it recommended the construction of a water main. The supply network was completed in 1872, although most of the 750 public and private fountains remained in use to such an extent that the threat of another epidemic led in 1884 to the closure of three-quarters of them. Fortunately, Cologne followed the example of other cities (London and Glasgow, Berlin and Wroclaw). The water from its fountains was analysed in 1862 and 1863, as was the water from the Rhine (in 1853, 1855 and 1859) that was being considered for distribution in large quantities.

In France as in Britain, the rapid strides made by the notion of public health during the 1830s had the virtue of integrating the question of water supply into a larger context which included the problem of housing, the cleanliness of towns, bodily hygiene, domestic habits, poverty and disease. On the East Coast of the United States, the situation was hardly any different, at least in the cities. The relationship of cause and effect between the quality of water and standard of health was not established; moreover, 'the effectiveness of the municipal authorities was restricted by the controversy on the transmission of disease that was raging between the proponents of the "miasma" theory and the supporters of the contagion theory.'[39] At the beginning of the nineteenth century, there was a considerable delay between the submission of proposals and the taking of the decision: twenty-one years in the case of Boston, thirty-one years for New York, twenty-nine for Strasbourg and sixty-one for Lyons! In the United States, this delay had been reduced to just one year at the end of the nineteenth century, by which time the old certainties had fortunately been rediscovered!

The water and health policies of local authorities were all the less positive since certain US newspapers had voiced the opinion during the cholera epidemics of 1832 and 1849 that Asiatic cholera was 'a disease of filth, of intemperance and of vice',[40] a sort of divine punishment. This is reflected in the words of the professor of medicine in Philadelphia who could not offer any other explanation; '"only the will of the Lord", he assured his students, "could account for the reappearance of a disease that had been quiescent for so many years." '[41] And early in June 1832, with a devotion strangely reminiscent of the consecration of the French nation at Sacré-Coeur after the defeat of 1870 and the Paris Commune, 'General Taylor recommended that the first Friday in August be observed as a day of fasting, prayer and humiliation. Mayors and governors immediately seconded the President's laudable injunction.'[42]

From the end of the eighteenth century in Britain and Germany, in France from 1802 onwards, as well as in several towns in New England, departments of health were established which, despite successive cholera epidemics, were rapidly engulfed by apathy for lack of funds and conviction. In the United States the Association for Health in Cities and then the American Public Health Association (1872) preceded the great French law on public hygiene of 15 February 1902. New associations and legislation encouraged the spread of health surveys and statistics and led to the organization of local and regional services which aimed to improve the general standard of health, and in particular the quality of water. This was a fine, almost premonitory idea!

Water and typhoid   It was also realized that typhoid, nicknamed 'the dirty hands disease' by the biologist Henri Vincent, was likely to be transmitted from person to person simply by handling the excrement of sufferers. This is what Professor Charles Achard said in a lecture given in 1929:

At the clinic in the Beaujon hospital [Paris] I saw three nurses who had not been vaccinated go down with typhoid fever while tending the sick. Twenty-five years ago, at the Tenon hospital, I saw a patient who had been given small tasks to perform in the laboratory at the height of an epidemic become infected from carrying patients' bedpans without taking any precautions and not without occasionally spilling a little of the contents. At the same time, my houseman also caught the disease while conducting an autopsy on a typhoid victim by taking urine in a pipette from the bladder: the cotton-wool stopper was sucked into his mouth, and with it a little urine; two weeks later he himself went down with the disease.[43]

This type of infection was not restricted to hospitals or to nurses and doctors, though it was under the nails of nurses that P. Carnot and Weill-Hallé found the highest incidence of Ebert's bacillus in 1915. Direct contamination of this kind also occurred in families, where the same absence of hygiene was by no means unusual; this can be clearly seen when several members of the same family become infected one after the other after having been in contact with a sick member of the family.

Propagation of this kind cannot be attributed solely to the patients themselves. Disease can also be spread by people harbouring an infectious microbe before the appearance of the first clinical signs, by convalescent carriers, by people who have recovered from infection without completely eliminating the bacteria (the bacillus may persist for years after the illness) or by healthy carriers, as observed for the first time at the Val-de-Grâce hospital by Remlinger and Sonweider in 1897. However, typhoid fever is just as frequently transmitted directly through food, linen and, above all, through drinking water, which is the principal vehicle for Ebert's bacillus, even in cities. Although it was suspected as early as 1823 and again in 1854, it was not until the very end of the nineteenth century that its role was highlighted, notably by the work of Drs Brouardel and Thoinot.[44]

Two types of evidence were used to confirm its typhogenic action: that derived from simple observation of the facts and other evidence derived from extensive examination of the suspected water. It was observation of the spread of epidemics that made it possible rapidly to establish the role of water in the transmission of typhoid. According to Brouardel and Thoinot, three types of cases had to be distinguished. The first involved the appearance of typhoid in direct topographical relationship to a particular water supply. The epidemic of 1886 in the Quimper *lycée* is a good example of this. There was not a single case of typhoid in Quimper when an epidemic broke out in the town's *lycée*: 34 cases were identified among the boarders, day boarders and staff. However, not one of the 155 day pupils was affected. This case made it possible to refute certain medical theories on the aetiology of typhoid which endured for a long time after Ebert's discovery of the bacillus in 1880.

The Quimper epidemic showed that the source of contamination was the water consumed in the *lycée*, which came from wells and tanks inside the school and which was the sole means by which the disease was transmitted. Further evidence for this was the fact that not a single case was declared in the town, except for a woman who was 'a great drinker of water' and who used her relationship with the school caretaker to fetch two or three litres of water each day from the caretaker's lodge.

This first type of observation, which underlines the link between social network and the geographical localization of the epidemic, was supplemented by a second type that showed that the incidence of typhoid fever varied in accordance with fluctuations in the drinking water supply. The quality curve rose or fell according to whether the composition of the water was altered by the addition of water from a different source. The illness began to disappear from the day that use of this typhogenic water ceased. The best-known example is that of Paris at the end of the nineteenth century. Under normal circumstances, the capital's water supply was provided by the Vanne and the Dhuys and the incidence of typhoid fever was fairly low. When the supply from either river was cut off for whatever reason, or a dry summer led to shortages in the normal supply, the water authorities added water from the Seine to the 'circuit'. Under these circumstances, a rise in the incidence of the disease was observed which did not fall again until the water supply returned to normal.

The third type of cases observed involved accidental contamination, triggering off a sudden massive epidemic that affected a lot of people in a very short time, sometimes simultaneously. At the end of July 1890, a soldier suffering from typhoid fever arrived in Trouville from Versailles. His excrement was thrown into the sunken cesspits that abounded in the Trouville area and which were veritable open manholes leading directly to the underground water level. Virtually continuous heavy rain then carried the excrement towards the deep wells and a break in the water main. The consequences were not long in making themselves felt; a massive epidemic rapidly developed (fifteen cases in one week in September 1890) among those using water from the deep wells and those who obtained their water from the municipal supply at a point downstream of the break.

A second type of proof to support the notion that contaminated water was 'the major pathogenic agent'[45] was also derived from analyses of water, and in particular from bacteriological analyses. Analysis of this kind was first carried out by Mörs in 1885 and then by Ivan Michaël in Dresden. The first bacteriological analysis in France was carried out by Chantemesse and Widal, who examined the water from a public street fountain in Ménilmontant and then the water in Pierrefonds after an epidemic of typhoid fever in 1886. These bacteriological investigations highlighted the mechanisms of contamination and spread considerable fear of these harmful 'impurities'. In their articles and political speeches, medical scientists did not fail to sound the alarm. In 1895 Brouardel and Thoinot wrote:

All drinking water can be contaminated by the typhoid bacillus. Shallow wells are the most easily accessible of all sources of drinking water. In the country, they tend to be close to dunghills on to which all excrement is thrown, or near streams and drains; they are subject to contamination from wash-houses where soiled linen is washed; or, even more simply, a whole range of rubbish is thrown on to the ground without any precautions being taken. In both town and country, it is normal to see cesspits and wells side by side, forming, as we have said, the two barrels of a gun.

Rivers have, for the most part, become open sewers.

Springs are also vulnerable, either because they are badly harnessed and thus not protected from contamination at the point of emergence, or because they emerge from ground that acts only as an imperfect filter, such as the fissured limestone of the Vanne basin in the area around Le Havre. If harmful matter is spread on this type of terrain it will penetrate far into the ground.

Another type of contamination is that which occurs not at the spring itself, but rather within the course of a distribution system: there is nothing more common than for the water supply in a town to be pure at the point of emergence, but for the distribution system within the town to be severed or ruptured, thus draining the contaminated urban soil.[46]

The rapid development of bacteriology, the science of the invisible, thus threw suspicion on the water that people consumed. It was not only the old-fashioned cleanliness, closely related to aesthetics and civility, that was considered necessary, but also modern, 'advanced' hygiene, involving regular washing of the hands, nails, genitals and underwear and cleanliness in the home. The discovery, three years apart, of the typhoid bacillus by Ebert (1880) and the cholera bacterium by Koch (in 1883) gave rise to a certain amount of questioning of Western practices in the area of public health. This was the beginning of a confusion between cleanliness and hygiene that is still prevalent. The rules of cleanliness are not necessarily the same as the rules of hygiene; the criteria for cleanliness are essentially cultural, whereas the basis of hygiene is purely scientific. If there is a difference between the social rituals of cleanliness and those created by the diffusion of hygiene, there is no need to cry shame or to parade outraged virtue. Each set of rituals belongs to a different system, even if they sometimes coincide.[47]

The practical questions

The evolution of knowledge about water gave rise to a certain number of practical questions.

The estimation of needs

Between 1760 and 1850, the predictions for water consumption drawn up by water engineers experienced rapid growth. Together with other European countries, France had entered the 'water devouring' phase of its history.

In about 1760, the mathematician Deparcieux put the water consumption per day and per Parisian at 20 litres. Many people at the time thought this figure exaggerated; one such was Mirabeau, who was to be involved in the launching of the first French capitalist enterprise set up to produce and sell water and who put the average daily consumption at 10 litres. In 1802, a certain Monsieur Bruyère, an inspector in the Highways Department, having ascertained that the average consumption per head did not exceed 5 litres per day in Paris, fixed the 'needs' of the average Parisian at 7 litres. In 1817, Prony, who had been entrusted by Napoleon with the task of draining the Pontine Marshes, stood by this figure of 7 litres. In Britain, on the other hand, this minimum daily requirement was put at 20 litres, and even as high as 30 litres by those who considered street cleaning and the washing of linen as vital progress. In accordance with the British hygienist model, Rambuteau and d'Aubuisson, the two municipal officials of Paris and Toulouse respectively, considered 20 litres to be very necessary. Some years later, in 1846, this figure was easily multiplied by a factor of six. A bylaw passed by the city of Paris put the domestic water requirement at more than 100 litres per day; 75 litres per day were said to be required per head of cattle or draught animal, 3 litres per square metre of vegetable garden, 20 litres for washing a carriage and 1.5 litres per inhabitant for cleaning the streets.[48] In Nevers, according to a scale issued on 31 May 1857 – which remained in force until 1909 – daily consumption was estimated by the concessionary company at 200 litres per household, to which were added 40 litres for each carriage, 60 litres per horse and 50 litres for gardens of less than 400 square metres.[49] Thus between 1760 and 1900, the overall estimation of 'needs' evolved considerably, from a few litres to several hundred litres per inhabitant per day. Forecasts of future needs no longer made a distinction between drinking water and water for domestic use, since the highly placed officials whose job it was to make these calculations had gradually accepted the notion of a uniform supply.

Line or network?

As the need for water increased, the water supply system in towns and cities was quickly to be called into question. At the beginning of the nineteenth century two rival techniques were competing with each other. The 'line' system, inherited from antiquity and given a new lease of life under the *ancien régime*, carried the water directly to a few specific locations which were supplied by branches that were totally independent of each other. The highest point was chosen as the centre, and the water pipe ended in a monumental public fountain 'reserved for the needs of the people', while 'a thin trickle of water' supplied 'the fountain of the privileged concessionaire by day and by night'.[50] Thus the water flowed continuously but not very plentifully, and the supply pipes did not constitute a coherent whole, let alone a network. A site was chosen for the monumental fountain and then the 'line' was constructed from that point in pursuit of a suitable spring.

A network, on the other hand, was intended to supply as many people as possible with a relatively copious flow of water. The hierarchical design of the network was calculated on the basis of the estimated requirements at the various 'terminals': Wallace fountains, public fountains, public toilets or fire hydrants.

These two methods of supplying water contrasted strongly with each other, particularly from a technical point of view. The technology of the 'line' system was simple and had been tried and tested since the work of the French physician Mariotte in the seventeenth century. On the other hand, the technology of the network system was much more complex, particularly with respect to calculating the dimensions of the pipes, and remained very rough and ready until Darcy drew up calculation tables in about 1855.[51]

The network method gradually came to dominate, but progress was slow, since the technical problems were further complicated by financial arguments. A supply system based on the 'line' method required a minimum number of pipes of modest diameter, whereas a network increased the length of piping required as well as the number of branches and offshoots, and because large volumes of water were being carried, very strong pipes had to be used. The case of Dijon provides a good example. Fourteen years after having opted for a 'line' system with a total of six monumental fountains costed at 60,000 francs, the city decided to install a network system at a cost of 1,250,000 francs, or twenty times that of the initial water supply.

The purification of water

Water, and in particular river water for supplying to towns, used to be free and available in unlimited quantities. In the course of the eighteenth century this situation began to show signs of change. A small, well-off and enlightened section of the population began to baulk at consuming this water, now suspect, which came from an unknown source or was delivered by porters.

Not long before the Revolution, private companies were set up on the banks of the Seine in Paris to filter the river water through sand and gravel. One of these 'factories' was established in 1771 at Port-à-l'Anglais in Vitry, and another at the end of the Ile Saint-Louis. Although output was still low (hardly 12 cubic metres per day), the water from the Seine lost one-third of its particles.

At the beginning of the nineteenth century, other processes were developed, particularly at the prompting of the navy, which had for a long time been grappling with the problem of conserving drinking water on board ship. Ground charcoal — either animal or vegetable — took the place of natural filters. The scientific academies supported these processes, notably the Academy of Saint Petersburg (1800), and there was a proliferation of learned articles in the specialist journals (in France, *Les Annales de Chimie* and *Le Journal de Pharmacie*). The chemist Berthollet proposed that the inside of the water barrels should be coated with charcoal. Others showed that even the filthiest water in Paris could be made 'excellent' for drinking by being filtered through a thick layer of charcoal. The testimony of members of the National Institute of France ensured that this process attracted a good deal of publicity.

In 1806, the first clarification plant was established on the Quai des Célestins in Paris. The filters were formed from a layer of gravel, a robust layer of charcoal mixed with fine sand and then a second layer of gravel. The water then returned to lead containers fitted with sponges which were changed every two or three hours. In 1823, other purification plants using the same process were set up in Rouen and Toulouse. However, the problem with charcoal filters was that they required frequent renewal. The engineer Genieys estimated in 1835 that, in order to obtain an output of 150 cubic metres per day, the minimum surface area required for the filters would be 7,000 square metres. The other major problem was the fouling of the filters. In the case of a private water supply, it was necessary to wash the gravel and replace the sand to a thickness that gave rise to a great deal of discussion. In 1857, the scientist Darcy concluded the debate. Having conducted a number of experiments in Dijon and supported them with results obtained in British towns, he advised that the surface layer of sand should be renewed every month to a depth of 2 centimetres. However, carbon filters were gradually abandoned after 1840, since it had been discovered that the charcoal was active only the first few hours of use and that it could even become harmful as a result of absorbing the decomposed oxygen.

Research led to the use of alum, which accelerated flocculation and facilitated the decanting process. Arago was an ardent supporter of this. By about 1848, another process had been developed which used filters of highly compressed wool. However, engineers eventually finished by recommending natural filtration, which limited the amount of handling and manpower required, even if it was slow and had to be carried out over large areas of the river bank close to the town or city.

After 1850, it was virtually impossible to draw domestic water directly from rivers, as the water carriers had been doing. River water, which since the Renaissance had been considered the best, was abandoned for a time. It was assumed that such water, even after filtration, could contain microscopic animals or 'putrid germs' and cause epidemics. Spring water was preferred; if possible, this water was conveyed by gravity from neighbouring hills through pipelines to the town. Some towns preferred to seek new springs to be tapped, while others used chlorine, alum or ozone to purify river water.

Artesian wells

Until the nineteenth century, it was hoped that artesian wells would make it possible to find new sources of water at low cost; however, technical problems hampered the use of this technique. Artesian wells are so called because it was in the old French province of Artois that the technique of sinking deep wells through impermeable strata emerged during the thirteenth century. However, it was not until the beginning of the eighteenth century that it began to be recognized as a technological innovation. Moreover, it was during this period that similar experiments were taking place in Lower Austria and Northern Italy.

Once the technique had been fully developed, artesian wells, by means of which the water gushing forth from the hole drilled through the impermeable strata into water-bearing strata could be collected, were in theory the most economical method of feeding water by gravity to a town.

However, the quality of the water obtained in this way was often questioned, as were the means by which it was procured. There were only two significant attempts to exploit this technique. The first was at Beauvais in 1745, and was carried out at the instigation of the mayor, after the town had already tried in vain to establish a supply of spring water; the other one was carried out at Caen in 1782, on the instructions of the provincial administrator. Both these attempts ended in failure, particularly at Beauvais, where, despite a well being sunk to a depth of 30 metres, the water that gushed forth did not reach the surface and had to be pumped up.

Everyone ran up holding out their pots, some even their hats, in order to catch some of the water: having tasted it, they all departed grimacing. The water was extremely clear and limpid, but bitter, acrid even; in a word, it tasted of minerals. Jacques Auxcousteaux, the mayor, who had hoped to have his name immortalized by this fountain, was desolated. . . . To complete this memorable failure, the well was blocked off.[52]

During the Restoration, the artesian well technique inspired more enthusiasm. It became well known through Diderot's *Encyclopédie*, the dictionaries of physics and the statistical yearbooks produced by the departments under the Empire, and to politicians and scientists it offered considerable advantages. The artesian wells in the north, for example, had a consistently high output, irrespective of the amount of rainfall, and their constant high pressure made pumping unnecessary. They also solved one of the problems of the expanding industrial sector, in that they offered a plentiful and constant supply of water at low cost. Competitions were organized and prizes awarded. Between 1818 and 1827, the Association for the Development of National Industry tried in this way to spread and perfect the artesian well technique. All this competition resulted in lower drill prices, improved performance and the drilling of numerous wells in several departments in the Paris basin. Artesian drills became endowed with some of the power attributed to the 'staffs' of prophets and saints. The following extract is taken from the inaugural speech given by the Minister of Agriculture to the French Royal Agricultural Society in 1829:

A skillfully handled drill entreats the ground to deliver up the spring that it is concealing and from which it cannot benefit. Nothing, neither space, resistance, fatigue, nor, most discouraging of all, doubt, can repulse their efforts. The drill quickens, it fights, it persists in its quest, and bubbling water gushes forth in pursuit of the departing metal, bringing with it life and fertility.[53]

Between 1820 and 1840, the number of artesian wells multiplied: twelve in Paris, eight in Saint-Denis, three in Mulhouse. Others were successfully drilled at Le Havre, Tours, Strasbourg, La Rochelle and Perpignan. Situated at a modest depth of about 30 metres, this pure, limpid artesian water could be obtained at a much lower cost than that of installing a supply network taking water from a river or a spring that then often had to be filtered and pumped. However, there were failures, notably at Epernay, Reims, Rouen and Mulhouse, and ever-increasing 'needs' could not be fully met by water from artesian wells. Thus after 1850, only a very small number of artesian well drilling companies were still operating in France. Artesian wells inspired much hope, but after several short-lived successes and in the face of competition from the network system, they fell into disuse, except in the minds of some local dignitaries who were ill informed about the latest developments in geology and water engineering; in any case, they were soon to be marginalized by technological advances.

The lifting of water

Once the hopes inspired by artesian wells had been dashed, attention had to be turned to the problem of raising and storing water. One of the main questions was how to convey water to the highest point of the town in order to distribute it by gravity. Aqueducts, much admired by Napoleon I and a whole generation of engineers at the beginning of the nineteenth century, had since fallen out of favour. On the other hand, using the energy generated by the current of the Seine to raise water seemed a novel solution, but it was not without its opponents.

Noisy, and sometimes even dangerous, the steam engine was far from receiving universal approbation: riverside residents were afraid of it and the mayors of the parishes affected agreed 'to wait until experience has led to improvements in steam engines, which seem to leave much to be desired'.[54] The engines consumed vast quantities of coal – an abundant commodity in Britain although still scarce in France around 1830 – were costly to buy (several thousands of gold francs) and gave only a modest output (0.3). Moreover, provision also had to be made for possible breakdowns, which would cut off the city's water supply for several days. As a result, efforts were made to increase the storage capacity of reservoirs, since water towers had proved inadequate. Attention turned to the construction of underground reservoirs, the capacity of which could reach several thousand cubic metres of water.

The pipes

Piping was the final area which experienced considerable technical advance. At the end of the eighteenth century, cast iron began to replace lead in the construction of urban water pipes. The cause of this change was the development of the metal industry, which led to very considerable reductions in the price of cast iron. The prices of lead and cast iron were virtually the same in the eighteenth century, but followed divergent paths between 1800 and 1850. Cast iron was half the price of lead in 1825, and only a quarter the price by 1850; in the case of France, this was attributable largely to the fall in the cost of transport.

However, the preference for cast iron was not only due to the industrial economy or the respective market prices of the two metals. There is no doubt that two traditions were in conflict: that of the fountain-makers, who preferred lead, which was the traditional material for fountains, and that of the engineers, trained in the prestigious engineering schools, who – confident of their own knowledge and taking the British example as a model – saw cast iron (and iron in general) as the material of the future. As the water supply networks were gradually installed, cast iron began to replace lead, if only because the influence of the fountain-makers on water supply projects waned as they lost their position as prime contractors.

In fact, between 1815 and 1830, the engineers succeeded in imposing their standards. Until then, manufacturers sold their pipes by weight and made them of a thickness equivalent to one-twelfth of the internal diameter. However, three innovations changed the standards then in force. Firstly, Genieys, who was responsible for the water supply network in Paris, tested the mechanical resistance of the pipes and, as a result, changed the way in which the thickness of the pipe was calculated. For those of large diameter, he estimated that a thickness of 7 per cent of the diameter was sufficient.

The second innovation was the introduction in 1820 of a new system of vertical casting in place of the old horizontal system. As a result, the thickness of the pipes could be more uniform and Genieys reduced it to 3 per cent of the diameter. These two innovations, when adopted by the manufacturers, reduced the weight of the pipes by half and doubled their length from 1.2 metres to between 2 and 2.5 metres.

The third innovation was the adoption of the British system of tendering. Henceforth, prices were not governed by weight, but rather a price per metre was agreed. This system was very quickly adopted by the prime contractors. It reduced the weight and average price of piping by one third. The end result was that in the twenty years between 1810 and 1830 the cost of cast-iron piping had been reduced by a factor of twenty.

Although cast iron had many advantages, such as easy handling and mechanical resistance, it also had its disadvantages, including frequent breaks, internal roughness which contributed to losses of pressure and a working life which failed to live up to expectations.[55] However, from

1830 onwards, new reaming techniques reduced the roughness, while at the same time the first bitumen-coated pipes came onto the market. Cast iron won hands down.[56].

Excretion and sanitation

The old system

The manufacture of reliable pipes was only one of the two aspects of improving the sanitation of a district or an entire city. As soon as a supply of pure water had been provided, it became necessary to find a means of getting rid of it once it had been used, since it was then full of waste matter and impurities. However, awareness of the risks of faecal contamination is a recent phenomenon which goes back only to Pasteur's time. Before that, until the middle of the nineteenth century, a minority of doctors but probably the majority of society at large believed in the beneficial properties of dirt and waste. Just before the outbreak of the French Revolution in 1789, butchers

attributed the good health that they generally enjoyed to breathing in the odours given off by the blood, fat and entrails of the animals that they slaughtered. In 1832, workers at the appalling refuse dump were still convinced that the fumes given off by excrement and other waste matter were beneficial to their health. . . . Peasant farmers persisted in keeping the indispensable dunghill just outside their doors. In Paris, the rag-and-bone men opposed the measures proposed by the city authorities. In 1832 they triggered off genuine riots against the decisions of the prefecture of police in its attempt to hasten the removal of mire and filth; they decided to use force in order to retain their rubbish heaps.[57].

Some at least of the enlightened elite of the eighteenth and early nineteenth centuries shared these beliefs, to the point of celebrating the therapeutic qualities of filth and excrement. Thus the *Encyclopédie Méthodique* reported quite seriously the official practice of opening up all the cesspits in London, a measure introduced in the city in the second half of the eighteenth century in order to banish the plague by means of the ensuing smell of excrement. In 1852, the great French hygienist Parent-Duchâtelet was still praising the therapeutic qualities of filth: 'in his view, they explain the good health of gut-dressers and sewermen';[58] some forms of consumption and rheumatism were even said to be cured in this way.

Fifty years later there had been a complete reversal of opinion. Scientists were doing battle aginst filth and excrement. Dr Calmette wrote:

All living beings from microbes to men produce excrement, which is the residue from their food and vital functions. Any accumulation of this residue soon begins to threaten their existence [. . .]. The higher animals and man would perish if they had to live among their own excreta.[59]

At almost the same time, the scholar Alfred Franklin, who had devoted a book to the private life of Parisians between the twelfth and eighteenth centuries, apologised as follows before broaching this subject: 'Although it pains me to do this, I must now broach an unappealing subject that is often neglected and which I too considered ignoring.'[60]

Sanitation

The early decades of the industrial revolution provide an explanation for this reversal. The simple (sic) notion that dirt must be considered a source of disease, and conversely, that collective and individual cleanliness was essential to the prevention of disease had gained ground. The reticence shown in both Europe and the USA during the Enlightenment had given way to the certainties advanced by microbiologists.

In 1842, Edwin Chadwick had focused attention on the deplorable sanitary conditions that prevailed among the working classes in Britain. Many large cities situated on small rivers had rapidly become 'poisoned' by human and industrial effluent. In London, the stench was so bad that, during the summer of 1859, the courts could not sit and sheets soaked in chloride of lime had to be hung from the windows of the committee rooms in the Houses of Parliament. In 1868, the committee studying river pollution 'received a letter, the writer of which considered it useful to point out that he had written it not with ink but with water from a river in Yorkshire!'[61]

Under the terms of the Metropolitan Water Act of 1871, the city authorities were given responsibility for ensuring that the water supplied by the private companies was fit for consumption. A water examiner and an auditor were appointed to inspect the water, a task which, in the public interest, was made subject to supervision by Parliament. Since 1865 a royal commission had been investigating the extent of pollution in the Thames; its final report was submitted in 1874 and legislation was introduced from 1876 onwards.

The members of the committee were aware of their duties as hygienists and were very much up-to-date with the latest scientific developments. Both their proof and their conclusions were closely argued. 'It is a widely spread custom, both in towns and villages, to drink either the water of rivers into which the excrements of a man are discharged, or the water from shallow wells which are largely fed by soakage from middens, sewers or cesspools. Thus vast multitudes of the population are daily exposed to the risk of infection from typhoidal discharges, and periodically to that from cholera dejections.'[62] As a consequence, not only the supply of pure water was regulated, but also the sanitation of towns and cities. The discharging of London water closets into water supply systems was, finally, banned!

The progress made in medical science, the social usefulness of which was now being acknowledged, meant that fresh support was gained for the notion of separating 'healthy' from 'unhealthy' water. For the German chemist Liebig, 'in the case of soluble nitrogenous organic substances' it all came down to 'a chemical process which takes place in the presence of oxygen and which is transmitted like a contagion' to all the matter present. According to this chemist's theory, living matter could be reduced to mineral. Essentially, fermentation amounted to a simple process of oxidation which did not take place unless life had departed from the substance.[63]

For Pasteur, on the other hand, fermentation was a phenomenon linked to life. Although he encountered many obstacles, he was to win the day over the chemists' interpretations. From his study of brewer's yeast (1855–6) and of other types of ferment, he concluded that the latter were living organisms which could multiply in the absence of air (anaerobiosis); fermentation and putrefaction were both transformations caused by various microorganisms (yeasts, bacteria, moulds) which grew by using the matter and energy produced by the decomposition of organic substances.

For a long time, the German chemists, led by Liebig, questioned Pasteur's methods and results. Ferments and the substance extracted from them (zymase) gave doctors the opportunity to amend their explanations of disease. Some went, quite shamelessly, from biological analysis of the phenomenon of fermentation to interpretations of the nature and causes of diseases, many of which seemed, from then onwards, to be caused by 'ferments'.

For the great French hygienist Bouchardat, the cause was clearly understood. 'In hygiene,' he wrote in 1867, 'the meaning of the term "marsh" is much wider than in everyday language.' Any unclean water encouraged the development of emanations; fermentation in marshes caused or encouraged the development of all the diseases described by medical science as 'zymotic': cholera, dysentry, typhoid fever, malaria, yellow fever. The ferment, whether vibrio or bacteria, fed on putrescible liquid. Bouchardat therefore recommended that hygienists should turn their attentions to the many different kinds of 'marsh' where fermentation took place, including mud, sewers, toilets, dunghills and cemeteries.

In the middle of the nineteenth century, nothing of the kind was yet taking place. The Munich school, personified by Max von Pettenkoffer, attacked the 'contagionists' and the theorists of 'miasmatism'. Thirty years after publishing his *Theory of Groundwater* (1854), which established a direct relationship between the level of ground and surface water and the intensity of certain epidemic diseases, the German hygienist declared to his counterparts in Alsace: 'For years I have been fighting the theories emanating from England that suggest that germs are propagated by water.'[64] Pasteur, Ebert, Koch: nothing made any difference! This refutation of Pasteur's theories had at least one commendable aspect. Like his British contemporaries Dr Southwood Smith and the engineer Edwin Chadwick, Pettenkoffer, who had perhaps read Friedrich Engels' *The Condition of the Working Classes in England*:

saw hygiene as part of a philosophy that embraced all of life's circumstances, involving not only pure air and water but also trees and flowers, which contribute to man's well-being and satisfy his aesthetic aspirations. He persuaded the city authorities in Munich to supply the city with clear water from mountains close at hand and to dilute waste in the Isar, downstream of the city. It was thus that the great clean-up of Munich began. . . . It subsequently became one of the healthiest cities in Europe, thanks to the efforts of this energetic hygienist who had not been influenced by the theory of germs in disease![65]

The debate on fermentation and contagion was coupled with a second, no less important argument. In their own jargon, the engineers of the day called it the 'combined/separate' debate. This controversy symbolically opposed two cities, two 'beacons' of Western civilization: London and Paris. London was to have a separate sewer network, while Paris was to install a combined system. The principle adopted in London was to provide separate drainage systems for runoff water on the one hand and sewage on the other, with the latter being piped into a completely sealed network. An additional supply of running water made up for any lack of flow associated with a still modest level of water consumption. Because of the lack of natural inclines in the area, 'powerful machines, located at certain points throughout the city [. . .], lifted the sewage to reservoirs from where it flowed downwards under pressure.'[66] Other pumps collected the effluent in reservoirs along the Thames so that it could be discharged at high tide into the river, in such a way that, diluted in the maximum possible volume of water, it could be carried away by the ebb tide.[67] The most powerful pumps could lift almost $5m^3/s$ to a height of 10 metres, and consumption in the system reached 44,000 tons per year.[68] Combined with an efficient system of collecting solid waste, mud and other rubbish by cart, this type of complex, scientific and expensive network was much admired by the future Napoleon III during his period of exile in London.

The system adopted by Haussmann and the engineers of the period was, however, different, if not directly opposed. Their solution was to construct a single network, known as a combined system. The size of the mains, which had to be enormous because of the volume of precipitation they had to cope with, meant that the network was monumental in scale. The rubbish collected by sweeping the streets entered the system through the manholes that linked the network to the surface and ventilated it. Logic required that all domestic waste be fed into the system without any prior separation. Even today, the argument is still not entirely resolved. At the time, the problems were obvious, since the flow in the system made it impossible for the network to cleanse itself. In times of drought, the flow of household water was not in itself sufficient to prevent separation, and the accumulated deposits began to ferment. By the end of the century, Parisians had already had several opportunities to complain of putrid smells. The worst of these incidents occurred during the summer of 1880; after a harsh winter during which the casks used to dispose of household sewage could not be flushed out, part of the contents of the 80,000 stationary cesspits was discharged into the mains. A dry spring and summer meant that the flow of water was not sufficient to carry away the sewage, which began to ferment. The incidence of infantile diarrhoea doubled and there was no hesitation in attributing this renewed outbreak to 'dangerous miasmas'.

The choice of system varied greatly, without taking into account the mixed systems. In about 1900, the combined system was preferred in large capital cities (Berlin, Brussels, Vienna, Rome, New York) and in many other large cities (Marseilles, Milan, Frankfurt am Main); in Europe at least, it was the commonest system.[69] During the second half of the nineteenth century, the separate system spread rapidly in Britain; it was also the commonest system in the United States,[70] as it was in the major Italian cities. Turin had a separate system consisting of two networks at right angles to each other. Naples installed a mixed system (double sewers in the lower and middle quarters of the city). Other cities, including Palermo, Catania and La Spezia, opted for separate systems. Of the 268 German cities that in 1900 had more than 15,000 inhabitants, 36 (including 8 with more than 100,000 inhabitants) had a complete combined system and no longer had any cesspits or casks, while about 80 had a separate system. In 1903 in France,[71] of a total of 616 towns with a population of more than 5,000, only 294 had a sewage disposal system, most of which were inefficient, while 65 had a complete combined mains drainage system. All French towns, even Paris, still had a certain number of cesspits, mobile soil tubs, etc. The worst cities in this respect were Marseilles, Nice, Nancy and Montpellier.

In what was probably an attempt at appeasement, the French doctor and civil engineer Ed. Imbeaux concluded in the report he presented to the International Congress on Hygiene in 1903: 'If properly installed, both the combined and the separate systems, as well as the mixed systems, satisfy the requirements of hygiene with respect to the rapid disposal of waste matter and water as well as of storm water.' There followed a series of somewhat complex technical assessments in support of this standpoint. Finally, an alarm was sounded, supported by a statistical outline of the sanitation system in Western towns and cities: 'very many towns and cities, particularly in France, either have no drainage systems, or poor or inadequate ones, and complete sanitation systems should be installed as a matter of urgency.'

The Parisian sewer system

Between 1850 and 1900, first in Paris and then in the other major urban areas in France, scientists, engineers and finally 'customers' adopted the combined sewer system. It involved – and this is still today the commonest system – 'draining off stormwater in order to prevent flooding and conveying sewage through a system of drains away from the city, either straight towards a natural outlet in the case of stormwater or towards a purification plant in the case of sewage'.[72]

In the course of the nineteenth century, the earlier sanitation systems in Paris underwent radical change. On the eve of the Revolution, stormwater and sewage had flowed together into 'streams' in the middle of the road. The flow of water followed the natural slope and, if it did not stagnate and flood the surrounding area, it found its way down a manhole leading into one of the several 'lines' of drains inherited from the past, most of which were open sewers which gave off noxious odours into the atmosphere. In this way, the water reached its natural outlet, the Seine.

Since 1350 people had been obliged to place waste in cesspits under their houses, which were then emptied by cesspit clearers; anybody unwise enough to insult them could be fined.[73] This system, which in Paris changed little between the fourteenth and eighteenth centuries, was modified in the nineteenth century and finally abolished by the 1894 act on the combined sewer system.

The first phase of the building of the Paris sewers began during the reign of Napoleon I with an exhaustive inventory of the system as it existed at that time. This was the period in which Bruneseau was given the new task of 'inspector of the Paris sewers'. In the words of Victor Hugo:

The sewer seemed endless and as awful as the *barathrum*[74] of ancient Athens. Not even the police had thought of exploring this decaying abyss. Who would have dared sound these uncharted and pitch-dark depths which inspired such terror? And yet one person did come forward. The sewers had their Christopher Columbus [. . .]. This man existed and his name was Bruneseau.[75]

Cross-section of an insanitary house in 1820. (Musée d'Hygiène de Paris). This model is intended to demonstrate devices inimical to the laws of hygiene: sealed cesspits and drain tanks for domestic waste water. (Photograph: © Boyer/Roger-Viollet.)

The surveys carried out from this time onwards not only located the old system but also discovered the routes taken by the effluent and served as a basis for evaluating the various construction procedures that were put forward.

After 1830, a sanitation system was installed in the northeastern districts of the capital. The water supply made available by the construction of the Canal de l'Ourcq facilitated street cleaning. The stream in the middle of the road gradually disappeared, to be replaced by two gutters on either side of the road. Public fountains were installed halfway up and on the top of hills. The manholes leading to the sewers were moved to the side of the road, the old drains were extended upstream and new ones built. Effluent was discharged either directly into the Seine or into the ring sewer, which flowed into the river at Chaillot.

The second phase of the history of the Parisian sewers is the most important one. Firstly, the rate of construction increased; the length of the system increased from 37 kilometres in 1824 to 130 in 1850 and to 560 kilometres in 1871. Secondly, the system began to be conceived in a different way; it was no longer simply a question of draining water and waste from the streets, but also of installing an underground collection system for sewage and rainwater. Henceforth, galleries protected the water supply pipes and were to reach a size that far exceeded the dimensions required for a simple drainage system. Finally, as a result of this massive urban sanitation programme, sites had to be chosen for the discharge of these enormous quantities of dirty water into the Seine, and at the same time the whole system had to be linked to a main sewer. Thus the principle of the unified, or combined, system was born.

There remained the problem of the cesspits. At the beginning of the fourteenth century they were emptied periodically and the waste used as agricultural fertilizer. However, according to the hygienists of the period (mainly doctors, architects and engineers), the cesspit system posed many problems. Because they were hardly ever impervious, the cesspits polluted neighbouring wells and liquid oozing out of them undermined the foundations of houses. Emptying them was a malodorous, arduous task and led sometimes to fatal accidents. Francisque Sarcey complained particularly bitterly about this in an article published in 1882:

Our forebears dug a large hole underneath their houses, which they called a cesspit. It was there that they carefully kept an accumulation of putrid filth and a hotbed of the most horrible pestilence. It was not emptied until it was full to overflowing. Very few changes have been made to this abominable system.[76]

However, as early as 1809, a prefectoral decree, which was superseded by a royal decree in 1819, attempted to make all cesspits impervious and to change the emptying system. Then, in 1818, a new type of cesspit

1870. Conveyed by manual workers, three members of the middle classes visit the Paris sewers, a source of astonishment and disgust.
(Photograph: © Hachette.)

was proposed: 'the odourless portable latrines of MM. Cazeneuve and Co'.

In order to prevent 'airborne contagion' and accidents caused by asphyxia and to reduce the stench caused by emptying the cesspits, a 'ventilator' was placed over the opening to the cesspit. With the same aim in mind, various chemical products were added to the waste matter in order to 'disinfect' it.

Finally, the long-established procedure of extracting the waste by means of buckets, baskets and soil tubs was replaced by a more modern system. Piston pumps and watertight containers replaced the carts which carried the teetering soil tubs to the refuse dump at Montfaucon, which it was now intended to close down.

A new site was opened at Bondy and the soil tubs were loaded on to barges. Then, since this method proved inadequate and because Montfaucon was still in use, a refuse disposal plant was built at the port of La Villette, from where the waste discharged by the cesspit clearers was conveyed through a pressure pipeline to Bondy. However, the plant at Bondy soon proved inadequate.

It was not until 1894 that it became compulsory in Paris to adopt the mains drainage system; this was twelve years after the city council had pronounced itself in favour of the new system. 'It was really only from 1905 onwards that we could consider the battle won.'[77]

The question, which had 'worn out five or six committees',[78] was indeed a complex one. It came up against long-established beliefs and affected the interests of two powerful groups: the cesspit clearers and the owners of blocks of flats. Thus the initiative, which came from the technical services department of the city of Paris, was not well received. Resistance was based on violent press campaigns which continued uninterrupted for twenty years, but opponents were unable to prevent the engineers from winning the day.

For all this, once the principle of the combined system had been established as the sole method of urban sanitation, the absence of any precise rules of calculation (for example for real flows) led to the construction of under-dimensioned schemes, usually as an economy measure, or, in contrast, of over-dimensioned systems; in both cases, the problem was severe enough to lead to malfunctioning of the system and excessively high costs. If a system was under-dimensioned, the infrastructure works had to be rebuilt periodically. When the system was over-dimensioned, it did not function properly: flows were excessively high and rapid and threatened to lead to catastrophe.

A more accurate understanding of the phemonena involved was obtained as a result of observations carried out on existing sewers forming part of completed systems; these measurements were used from 1920 onwards as a basis for drawing up more adequate calculation rules.[79]

Finding a solution for the problems involved in installing a sanitation system in Paris was a long process which led finally to the adoption of the combined sewer network as a basic principle. Thus it was that the utopia of Haussmann, with Poubelle the most famous of the prefects of the Seine, was realized:

The underground galleries, the vital organs of the great city, would function like those of the human body, without being exposed to the light of day; pure, fresh water, light and heat would circulate like the various fluids whose movement and maintenance are essential to life. Secretions would take place there mysteriously and would maintain public health without disturbing the orderly running of the city and without spoiling its external beauty.[80]

2

New objectives

Once water had been analysed by the scientific elites and tamed by the engineers, the conquest began, initially in towns and cities. The proliferation of fountains, public wash-houses and shower baths changed the urban landscape. The extension of the subterranean universe created a sort of invisible, upside-down city, bluish-green and evil smelling. Housing began to be planned and the water supply and drainage systems were improved, making it possible to provide running water in kitchens and bathrooms. Water gradually became an indispensable element both in public areas and private houses which since the eighteenth century, had become differentiated from each other, first of all in the upper reaches of society and then in the lower middle and middle classes. By the nineteenth century, daily life within dwellings was no longer characterized by excessively close proximity to children, servants and friends. Similarly, spaces were no longer used for a multiplicity of purposes. Work and private life were usually kept apart, at least in towns and cities. As a result, there were changes in the design of houses and in the use of particular rooms within the house. The 'privy' became a separate room. Attending to personal hygiene became a solitary activity that took place in a refuge that was often dark (intentionally so. . .). When the location of the bathtub in richer dwellings is indicated, it was in a set of bathrooms that was separated off from the other apartments that made up the block.[1] This change reflected changes in mentality, habits and the use of space that took place in the upper strata of society during the eighteenth century:

From the eighteenth century onwards, the family began to distance itself from society, to push it back to the edges of a private life that was constantly extending its boundaries. The design of houses reflected this new preoccupation with erecting barriers against the world. . . . It has been said that comfort dates back to this period; it was born at the same time as privacy, discretion and isolation and is one of the ways in which they manifested themselves. Beds were no longer placed just anywhere, but moved to a separate bedroom, fitted on either side of the alcove with cupboards and closets in which there appeared new instruments of hygiene.[2]

As a result, the demand for water for washing and, more generally, for personal hygiene increased considerably. A new period was emerging in succession to another, much older one. It was no longer a question, as it had been in the past, of washing 'only out of regard for others, and more particularly for those more highly placed in the social hierarchy'. At that time, now past, the motivation was essentially social and often reflected external constraints; when these constraints were absent or when social standing did not impose any obligation to wash frequently, cleanliness was limited to the minimum required to maintain personal wellbeing. This period stands in marked contrast to the present period in which 'children are made from a very early age to wash and clean themselves; as a result of this conditioning, washing oneself becomes a virtually automatic process'.[3] Children finish by 'feeling constrained to wash themselves even when nobody is about, even when nobody could punish them for having neglected their personal hygiene'.[4] As far as hygiene is concerned, the evolution of civilized habits had followed a very clear path: 'social relations changed in such a way that, in each individual, mutual constraints took on the character of internal constraints, leading to the formation of an increasingly more pronounced "super-ego".'

Cleaning public spaces and adorning them with fountains, providing each dwelling, each family, each individual with an abundant supply of clean water and draining away and treating waste water were all very concrete aspects of the conquest of water that had their origins in the implementation and rationalization of the notions of hygiene advocated by our near ancestors.

The public sphere

In the course of the nineteenth century, an increasing number of fountains and wash-houses, symbols of the conquest of water, were installed in public spaces in towns and cities.

These fountains retained their utilitarian function, which was obvious when they were simply water delivery pipes bored into a hollowed out trunk, but they also had an aesthetic aspect which emerged from the thirteenth century onwards, particularly in the cities. This was certainly the case in Paris and Lyons, with a circular stone basin, in the centre of which stood a pillar surmounted by a ball from which the water gushed forth.

Utilitarian and aesthetic at the same time, fountains were also sacred springs which 'bear witness to the role of water as a secular and religious symbol of youth';[5] in this guise, it appears in the paintings of the period, such as the famous *Mystical Lamb*.

This gushing water, bursting with power, was also heavy with symbols. Each period had left its mark on it. During the Renaissance, it was a pagan symbol, associated with the games of antiquity; as a monumental product imported from Italy, water had a role to play in the celebrations of the princes who adopted it. It was present at royal entrances, adorned the parks of Renaissance chateaux and drew inspiration from the renewal of interest in antiquity. It celebrated the rediscovery of nudity and the pleasures of the body; the development of suction and force pumps transformed water into an enchanting spectacle for the dazzled eyes of the court: 'In the seventeenth century, water, embodied by Versailles, regained a certain sobriety. Le Nôtre and Mansart subjected it to the laws of gravity and of adornment governed by geometry.'[6] The system developed by the hydraulic engineer Francini transformed the old marshy pond into a scene painting of water and stone in which, alternately, water jets triumphed over gravity, water flowed in perfect harmony with the naked bodies of the statues and the Basin of Apollo celebrated the Sun King. Ancient links between the utilitarian, the aesthetic and the symbolic were thus recreated, in a language that had been revitalized in favour of absolutism.

From Versailles to Sans-Souci via the Hofburg, water was gradually being 'gallicized'. In the eighteenth century, every town, however modest, felt obliged to erect a monumental fountain. These fountains generally retained their classical style, but gradually freed themselves from such restraint in favour of curves and counter-curves.

The 1850s were to see a change in this balance between the utilitarian and aesthetic functions of fountains. The former began to disappear as the domestic water supply system was extended and the impracticalities of fountains became more manifest. The decorative monumental fountains symbolized the triumph of water in public places, whether royal, imperial or Republican. At the Universal Exhibition of 1889, the magic of electricity brought about the union of water and fire: the illuminated Republican fountain replaced the fountains used in the royal celebrations of Versailles.

On the other hand, the sacred and symbolic functions of the spring were to be maintained into the twentieth century. In any case, all these functions are not mutually exclusive. From the 1840s, the long-established utilitarian function took on a hygienist aspect, while the aesthetic function was sustained because the middle classes, like the aristocracy, adored aquatic mirages and the reflection of the city – the beacon of civilization – in its fountains.

The symbolic aspect of the fountain, which usually indicated the vitality and power of the town or city, was also preserved. This was certainly the case at the beginning of the eighteenth century at Saint-Cézaire in Provence where, during a drastic water shortage, a small fortune was offered to water diviners able to discover a new spring. On the day that the fountain was inaugurated and blessed, there were tears of joy on everybody's cheeks.[7]

The sacred function proved to be equally deep rooted and capable of changing in accordance with technical innovations and cultural choices. Even today, particularly in Catholic countries, certain fertility rites still persist in which the fountain is placed under the protection of a local saint or of the Virgin Mary. Equally, the washing of clothes, seemingly a highly secular and extremely utilitarian task, for a long time retained a sacred aspect, reflected in the notion that the act of laundering, like taking a bath or spring cleaning a house, represented a return in time to an earlier period in which things were restored to their starting point; dirty clothes thus became clean again and life was restored where death had reigned. Once revitalized in this way, time started again from zero.[8]

A particular type of fountain, the washing fountain, also emerged in the course of the nineteenth century. These were monuments of moderate size, located in the centre of the main square of villages, or banished to a site behind the church or the town hall, between two large houses, or even built over the springs themselves on the outskirts of the village.[9] Some of them were built as early as the end of the eighteenth century, but most of them date from between 1820 and 1880, usually from the time at which the region reached its highest level of prosperity. This was also a period in which the hygienists' desire to control the sources of water merged with the urbanist aspirations of village dwellers.

In the case of the washing fountains dating from the eighteenth century, the drinking trough and the washing place were built on the centre line of the spring, and the little building erected over the spring for the purpose of drawing water had a roof which sloped in a manner reminiscent of a dome. At the beginning of the nineteenth century, the previously rectangular washing place became oval or octagonal.

As fountains proliferated, wash-houses began to appear in large numbers in public spaces. Before this development, and despite official prohibitions, washing was done anywhere, in buckets, in pools, by the side of a river or stream or in fountains. This situation underwent radical change in the nineteenth century as wash-houses built on dry land proliferated and replaced washing places situated along a water course or on river banks, as well as the floating wash-houses that had become established in Paris at the beginning of the century and which at the time had represented 'the first attempt to create a clearly defined area for the washing of clothes along the banks of the Seine'.[10]

Indeed in the eighteenth century, the need to divide up space was less urgent.

The crowded banks of the river draw considerable crowds and cause many accidents [. . .]. The water, which is essential to the life of Paris, attracts a large number of minor activities, Laundresses beat their clothes and put up their interminable washing lines; water carriers quickly fill their buckets before setting off up the terraces.' [1]

A brief glimpse into daily life along the banks of the Seine is afforded by the careful records kept by the city authorities of the rewards paid to the rescuers in the incident described below:

16 January 1779 – 10 o'clock in the morning – Joseph Colin, known as Champagne, a water carrier aged about 50 and living in Rue Jean Beausire, was standing, laden with his full buckets, on the water scoop at the head of the boats when he lost his footing and fell into the river; he was carried away struggling by the current. Four washerwomen, Marie-Jeanne Boissy, Marguerite Desgranges, Marie-Anne Vincent, all residents of the Rue Jean-Beausire, and Jeanne Gauché, of the Faubourg Saint-Antoine, hauled him out after he had been immersed for several minutes; he was carried to the guardhouse on the Isle Louvier where the sergeant had him undressed, cleaned and dried; he was then warmed up and given camphorated brandy and had no need of further assistance. After his comrades had brought his clothes, he was sent home.

To the women who hauled him out ....24 livres
To the sergeants and guards...................... 9 livres
To Mr Hoin for his attendance................... 4 livres
Expenditure on wine, sugar, laundry . 1 livre 10 sous[12]

Almost seventy years later, the places where it was permitted to wash clothes were more precisely delimited. In 1858, there were 171 washing places in Paris all of which were run by private entrepreneurs and frequented mainly by the working classes. Sixty-four floating wash-houses were anchored along the Seine between the Pont Marie and the Pont des Invalides, with a further seventeen on the Canal Saint-Martin. On land, there were ninety-one wash-houses and laundry rooms in the heart of Paris – in the rue de Bièvre – or just beyond the first line of boulevards, in the rue de Ménilmontant or the rue du faubourg Saint-Denis. These wash-houses had many disadvantages, including a lack of equipment and of places for drying clothes, and cramped, confined working conditions. This was attributable to the owners' desire to squeeze as much profit as possible from the space occupied by the wash-house. The authorities were relatively unconcerned by these basic problems; from the eighteenth century onwards, however, they did fear the moral dangers posed by the wash-houses. Professional washerwomen, who were a race apart from ordinary housewives and were separated from them in the Paris wash-houses, had loud, aggressive voices, quick tempers and a bad reputation. The bargees called them 'moorhens'. . . . They had to fight against fatigue, cold, damp and the occupational diseases to which they were exposed. During a visit to Nîmes in August 1844, Flora Tristan wrote:

There is only one wash-house for all the washerwomen, dyers and others, and what a wash-house it is! Imagine a hole, laughingly called a basin, hollowed out in the middle of a square (I do not know its name). This hole is about sixty feet wide, 100 feet long and forty feet deep. A two-runged ladder leads down into it. There are two washing boards which stretch the entire length of the basin but which are only one foot in width – now guess how these washing boards are constructed! Like any other washing board, you say. Ah! but that's just it! You see, they are constructed in exactly the opposite way to all other washing boards. In all other wash-houses, the stone on which the laundress washes her clothes slopes into the water so that she can scrub the clothes in the water. The laundress kneels or stands (as in the boats in Paris), and washes her clothes on the sloping stone. This is so simple that women in the country make themselves a washing place at the edge of a river or stream by laying down a stone in such a way that it slopes into the river; they then kneel behind it to do their washing. Well, in Nîmes, things are done the wrong way round. It isn't the clothes that are in the water, no, it's the woman washing them who stands up to her waist in water, while the clothes are out of the water and the laundress washes them on a stone sloping away from the water. Thus 300 to 400 laundresses in Nîmes are condemned to spend their life up to their waists in water, and in water, moreover, which is poisonous because it is full of soap, potash, washing soda, bleach, disinfectant, grease and all sorts of dyes, such as indigo, madder and saffron. In order to earn their daily bread, many women here are condemned to diseases of the womb, acute rheumatism, difficult pregnancies, miscarriages, indeed all imaginable afflictions! I ask you: has a more revolting atrocity ever been known, even in the most barbarous of countries, than that which is being perpetrated against the wretched washerwomen of Nîmes! A convict would not be condemned to suffer for just one week what these unfortunate women have had to endure for the last three hundred years since the wash-house was built![13]

After 1848, admiration for wash-houses of the British type began to grow. This was taken up by Napoleon III, not without a certain paternalism, and the technical excellence of the British model was much lauded. Delegations were sent to Britain; they were able to see the success of the wash-houses and praised the rapid and efficient procedures in the wash-houses of London and Liverpool. However, this enthusiasm did little to break down the hostility of the wash-house owners, who were not on the whole well disposed to the threat of possibly ruinous competition. However, the French emperor did take decisive action on one point: a bath and wash-house based on the British model was financed from the civil list. It was originally planned to build twelve such wash-houses, but only one was actually finished. It was built by two architectural engineers from London on part of the former site of the Temple ceded by the city of Paris, and was based on the British concept of combining a bath-house, which was profitable, with a wash-house,

which was loss-making.[14] Certain materials were imported from Britain,
notably Portland cement, which was used for the first time in France.
British steam engines and drying and ironing machines ensured that the
technical side of the operation ran smoothly. New methods, such as
steam washing, which had been in use since 1853 in military laundries,
warm-air drying, the use of hot water from a tap and ironing rooms,
complemented the new equipment and were intended to make the work
of washerwomen and housewives less arduous. A vapour extraction
system was even installed in order to maintain a healthy atmosphere in
which to work.

The new wash and bath-house was opened on 15 May 1855 and was
overwhelmingly successful. It was open twelve hours a day in order to
meet the demands of customers attracted by low prices and comfortable,
up-to-date facilities. However, financial difficulties rapidly accumulated;
building faults and a lack of day-to-day maintenance made it necessary
to carry out expensive repairs that could not be paid for from the meagre
profits made by the new venture because of the low prices and the lack of
space. The operators abandoned the British system of laundering and
returned to more profitable methods: a tub or vat for the washing, no
more hot water in the booths and internal alterations in order to increase
the number of places. This made the wash-house profitable while at the
same time destroying its original character.

In 1861 – six years after it had opened – the building was so
delapidated that the city authorities were able to make it subject to a
compulsory purchase order and to use the site to build the
administrative headquarters of the 3rd *arrondissement*. Thus the
experimental bath and wash-house, which was to have been a model of
hygiene and edification, was condemned. This failure and the
abandonment of the plan to build a further eleven such wash-houses can
be attributed to Haussmann's policies; the water supply projects
implemented by Belgrand and the building of apartment blocks with a
water supply on all floors were henceforth given greater priority than the
construction of public wash-houses for use by the lower classes,
particularly since urban development policy was forcing them out of the
old districts of the city. Moreover, the renting of these establishments,
which made deficit budgeting impossible, and the complaints of
housewives who said they were no longer allowed to speak, contributed
to the closing of the Temple wash and bath-house.

It should not be concluded from this that the large wash-boats and
the small family laundries had disappeared by about 1850–60. On the
contrary, 'the washing of the clothes of the middle classes occupies an
army of specialist washerwomen,' of whom there were 70,000 in Paris at
the end of the Second Empire.[15]

Armed with water and courage, washerwomen 'banish dirt' at the communal wash-house. (Painting by Paul Grégoire, Salon of 1909.
Photograph: © Roger-Viollet.)

Everywhere in the city one sees free, vigorous, lively, independent women delivering their washing and peddling rumours just like their male counterparts the water carriers. They are outspoken; there are long boats filled with washerwomen whose tongues wag as energetically as the paddles with which they beat the clothes.[16]

With their paddles in their hands, all these women were ready to strike, to engage in unlawful assembly and even to riot. They were immortalized by Zola in *L'Assomoir*, where he has them applauding and hurling abuse at Gervaise during her fight in the wash-house. Gervaise is a housewife, a working-class woman who, like her fellow washerwomen, goes to the wash-house when her other duties as wage-earner, housewife and mother leave her two or three hours of 'freedom'. The wash-house was a place given over to women's work and talk, a place of intense exchanges, of intimacy and solidarity; it gave rise to a new kind of sociability among women which until that time had been given expression only once a year during the great spring wash.

Despite some acknowledged failures, the government, in the name of public health, encouraged the establishment of these public wash-houses, both in the town and the country. In fact, the 'paid washerwomen, who were poor women, often widows, with chapped hands and aching backs' had been carrying on their trade for a long time, like the women factory workers, but because they washed clothes in the fountains, the soapy water found its way back to the spring or 'reached the drinking-trough saturated with soap and dirt from the clothes, so that the cattle had only unwholesome water to drink'.[17]

Even in the depths of rural France, where cleanliness was not a major concern, the number of public wash-houses increased greatly between 1820 and 1910. Each parish built its own wash-house; they were fairly rudimentary buildings, closed in on two or three sides and usually devoid of any aesthetic merit. Nevertheless, the parish wash-house 'where the wind blew violently in bad weather'[18] represented a significant step forward. As Daniel Halévy noted in 1910 with regard to Ygrande:

The municipal councillors have had a covered wash-house built, 18 metres in length and 8 metres in width. This will render a service to all housewives in the town, writes Emille Guillaumin, but more particularly to the poor women whose main source of income is to do other people's washing and who consequently wash clothes every day, from morning to evening and in all weather: in winter they sometimes had to suffer atrocious days.[19]

In the great traditional wash – which was to become a thing of the past with the introduction of iron boilers[20] from 1900 onwards – the washerwomen did not start to play their role until the second phase of scrubbing and rinsing that followed the first phase of washing:

The washing or boiling phase is slow, serious, long, almost solemn; it is a private, domestic task in which the slow chemistry of the embers gradually does its work. The scrubbing and rinsing stage that follows brings to the wash-house the band of washerwomen, with all their noise, activity and good humour.[21]

Until the arrival of washing machines, this second stage of scrubbing and rinsing clothes in the wash-house was still a part of the laundering process in some rural areas.[22] In Minot, the traditional washday continued to be used by peasants as an opportunity to show off their best clothes to the rest of the village. After 1960, the introduction of domestic washing machines turned what had previously been a public activity into a private, family affair.[23]

Both in the town and in the country, it is undeniable that the communal wash-house was a monument to the new cult of cleanliness.[24] Between 1870 and 1900, more than 200 wash-houses and laundry rooms were opened in Paris, and by 1909 there were 400 in all. Even though the unpopularity of the wash-house owners was an established fact, the institution was a flourishing and obviously profitable one. As a result, there was a well-regulated procedure for scrutinizing applications to open wash-houses. In 1858 in Bordeaux, 300 out of a total of 516 applications were rejected on the grounds of insalubriousness. In 1860, a number of plans were put forward for massive wash-houses that would each be able to accommodate a veritable 'army' of 200 to 800 washerwomen. From 1870 onwards, a national association of wash-house owners was established and rapidly became powerful. In Paris between 1870 and 1880, it was estimated that each of the city's inhabitants washed or had washed three kilos of clothes and other items per week. In the period between 1840 and 1870, plans for new apartment blocks usually included a communal wash-house.[25]

All the wash-houses of the time were filled not only with the smacking of the paddles as they beat the clothes but also with the noise of wagging tongues. The day had its own rhythm:

During the coffee break, travelling coffee sellers and confectioners would bring the women refreshments; during the lunch hour, the women would go to the canteen to eat the snacks they had brought with them or buy the inexpensive meal that was on the menu that day. They talked and looked after the small children that some of them had brought with them, putting them down to sleep in buckets that served as cots.[26]

During the lunch hour, hawkers would peddle their wares, followed sometimes by travelling photographers and strolling singers, many of whom earned their living by going from wash-house to wash-house.

What did the washerwomen talk about? Of the intimate life of the village as reflected in its washing. As the washing was being sorted in the wash-house frequented by Gervaise, Clémence brought into play her ferociously sharp tongue:

Thereupon the cheeky beggar had her say, some dirty word or other, over each article; she laid bare all the misfortunes and all the misconduct of the customers, joking over all the holes and all the stains that passed through her hands [. . .]. So, at every sorting, the whole neighbourhood of the Goutte-d'Or was taken to pieces at the shop.[27]

The water towers

If the proliferation of fountains and wash-houses in the nineteenth century changed the character of public spaces, the same was also true of water towers which, unlike aqueducts, were a new phenomenon. Although aqueducts continued to be used[28] and restored,[29] sometimes only partially,[30] they were manifestly not one of the characteristic structures of the century.

The water tower, on the other hand, was a new phenomenon. In France, the construction of water towers was closely supervised by a government agency, since they formed one of the key elements in the water supply network. It took hydraulic engineers in Toulouse and Marseilles three years to decide on the location and capacity of water towers in the area. In view of the head of water, the storage tank had to be built on stable ground not subject to flooding and in close proximity to a water source. It had to house the pumps and the rising main and, since it was located at the head of the supply network, it had to be close to the town or city that it supplied. The storage depot for pipes and fuel, the stables that were needed when horses provided the power supply, and the tower keeper's house, meant that the site required was fairly extensive. The site chosen for Toulouse was at the far end of the bridge over the Garonne, while the site in Chartres was a platform near the ramparts. In both cities, the storage tank was sited 35 metres above the ground, which was the height required to convey the water to the highest point of the town.

Before the nineteenth century, there were only a few water towers in existence, notably the wonderful Toulouse water tower, which dates from the seventeenth century, the marvellous monument on the Promenade du Peyrou at Montpellier, which fed the city's fountains, and the Mannheim water tower in Germany. It was not until the second half of the nineteenth century that the need for such elevated water reservoirs began to be generally accepted. Until then, efforts had been concentrated on line and fountain-based supply systems. The beginning of the twentieth century saw the development of a modern combined system which provided individual users with their own supply. Water towers then proliferated, both in the town and the country. Similar in shape and size to belfries, steeples and towers 'the water towers, those concrete warts on legs, disfigure the countryside and, along with railway stations, gasholders and grain silos, occupy a special place in the museum of architectural horrors.'[31]

Indeed, until the competition organized in 1939 by the Committee for Hygiene and Water, 'water towers (were) strictly functional and built with no regard whatsoever for aesthetics in accordance solely with technical constraints such as water conservation, sufficient height to obtain the required pressure, a predetermined sheet of water in order to keep variations in this pressure within reasonable limits and economy of operation linked to the thickness of the walls.'[32] With very few exceptions, the water towers built in the years between 1830 and 1940 had absolutely no artistic pretensions. In France, it was only with the innovations of Auguste Perret at Saclay (1942) or those of the painter Vasarely that any aesthetic concerns began to surface.

Fountains and politics

Just as Britain has its fountains and spas (Trafalgar Square, Bath), so the West has its monumental fountains, the beauty of which still captivates the imagination. In about 1850, there were twenty-six fountains of this kind in Paris, all of which used about a considerable 300 m³ of water per day. Nor was Brussels any exception to this 'rule'. Among others, the Belgian capital could boast the fountain in the Bassin du Parc, the Fontaine Rouppe and the two fountains in the Place des Martyrs. As for the famous 'Manneken Pis', there was nothing monumental about it and the statue dated back to 1619. The grace of this 'natural act' was a symbol of Belgian vitality which the code of decency would have prohibited two or three centuries later.

The Fountain of Neptune in Berlin, which adorns the square in front of the town hall (now in East Berlin), was designed by Reinhold Begas and erected in 1891; its allusions to Roman mythology are typical of the period. The Märchenbrunnen (or 'Fairy Tale Fountain') is more Germanic in style, reflecting another past and celebrating a deep-rooted identity.

'A jewel in the crown of official art, the nineteenth century fountain, which was hardly innovatory',[33] was subordinated to the sculpture for

One of the Wallace fountains in Paris (late nineteenth century). About a hundred of them provided water to slake the thirst of passers-by.
(Photograph: © D.R.)

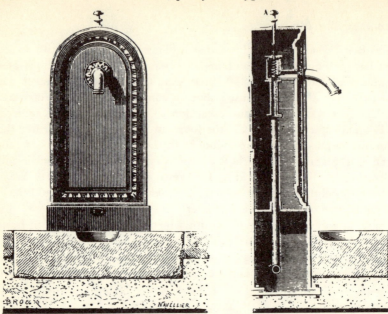

Front view and cross-section of a street fountain. Before each house had its own supply, water was supplied in this way to neighbourhoods, sometimes as early as 1830 and on a more general basis after 1860.
(Photograph: © D.R.)

which it provided a setting. The justification for building the fountains was the allegorical messages that they conveyed. One of these was based on the rivers associated with the history of the city or, sometimes, of France, a very traditional and already much-used idea that was given a new lease of life. In the Place des Terreaux in Lyons, the traditional sea chariot symbolized the Saône and its tributaries (1892); in Paris, the four rivers of the fountain in the Square Louvois symbolized France itself.

In the nineteenth century, the advances made in establishing a water supply system freed fountains from their utilitarian function. As purveyors of allegorical messages, fountains were called upon to play a variety of roles: they served to commemorate peace (the Fontaine Saint-Michel in Paris, by Daviou), victory (the Fontaine du Châtelet in Paris, by Bralle) or the centenary of the Republic (the fountain in Toulouse designed by the Allar brothers).

After 1870, in addition to these allegorical fountains, there appeared many monuments consecrating the Republic, usually in the form of a woman wearing a Phrygian cap as a symbol of liberty. Virtually all of them were built in a period in which running water was still a rarity, when clothes were often washed in fountains and animals slaked their thirst at drinking-troughs. The whole village community thus passed in front of these Mariannes. Indeed, these monuments were not very common in the major cities such as Paris. The statue erected in the capital in 1883 in the former Place du Château d'Eau, now the Place de la République, is surrounded by a large fountain which here plays only a decorative role.[34]

On the other hand, in the Republican villages of Lower Provence or Languedoc, the simple geometric design (column with capital, sphere, pyramid) used earlier was replaced, from the middle of the nineteenth century onwards, by a monument decorated with figures. Nevertheless, Marianne did not have a monopoly over this role. The Virgin Mary, saints, agriculture represented by Ceres or Bacchus, the bust of a benefactor or of a good mayor were all rivals.

The simplest explanation for this association between water and the Republic is a financial one. It was cheaper to perch Marianne on the fountain in the square rather than go to the expense of a pedestal. However, there is also an ideological explanation: the fountain and the statue were two aspects of progress. At Eygalière, two inscriptions can be seen, in perfect symmetry with each other: 'French Republic – Liberty, Equality and Fraternity' and 'Water supply, 1900, E. Bouer, Mayor'. Finally, and less positively, it is difficult to refrain from reflecting on a symbolic interpretation: Marianne, the new female divinity with which the Republican pantheon had just been reinforced,

Allegorical fountain, Nancy, Place Royale. A sea deity, the four rivers of the region and agricultural symbols (wheat and grapes) bear witness to the prosperity of a provincial capital.
(Photograph: © D.R.)

was associated with running water, the traditional image of life and prosperity.

The monuments which best symbolized the association between water and local authorities in France were the combined town halls and wash-houses that were constructed in the nineteenth century. These were mixed buildings that were reminiscent of the combined covered markets and town houses of medieval towns and the town halls-cum-schools that were built in France during the Third Republic; the wash-house was on the ground floor, with the town hall on the first floor. In addition to these two communal functions, the building often contained the village school, and the communal oven was often nearby. Water, which had hitherto been the preserve of aristocrats and the rich middle classes, was gradually democratized and came under the control of local authorities. Thus the cult of cleanliness as the huntress of miasmas and death and the pleasure of hearing and seeing a free water supply flow provided the village with a centre for its social life which was secular (in opposition to the place of the church), salubrious and brought to life by the washerwomen, by the toing and froing of horses coming to drink at the trough and by men hurrying to the town hall. The women's domain was the ground floor and the intimate life reflected in the washing they brought there, while politics and public office remained the sphere of men.

The private sphere

The development of knowledge about water was not restricted to the public sphere, either in the town or the country. After the time required for the individualization of supply and drainage systems, it also began to manifest itself in the private sphere, whether it was a question of sanitary equipment (washbasins, baths, bidets, toilets) or of toilet articles (razors[35] and toothbrushes, for example).

Washbasins

The term '*lavabo*' (the French word for washbasin) is borrowed from the language of ecclesiastical liturgy, in which it denotes the receptacle in which the celebrant washes his or her hands while reciting some verses from Psalm XXVI:

Lavabo inter innocentes manus meas et circumdabo altare tuum.

> (I will wash mine hands in innocency:
> So will I compass thine altar, O Lord)

This cleansing ritual, which is shared by many religions, is carried out before the oblation and is a reminder that mass is a sacred meal. It represents the survival into the present of much older, pagan traditions that were adopted by Christianity.

Similarly, in secular or material life, the frequent washing of parts of the body is a practice that has always been maintained, even in the Middle Ages. It used to be the custom to wash one's hands on rising and before and after meals. In the latter case, this was a necessity because, until the seventeenth century, forks did not exist.

In the well-to-do classes, servants would go round the table with a basin, an ewer and a washstand. The washstand, a distant ancestor of the modern washbasin, could be either fixed or portable; in the thirteenth and fourteenth centuries, it consisted of small basins, usually made of enamelled copper, one of which was used for pouring and was decorated with a gargoyle, while the other one was used for washing.

However, it was a second type of washstand, in common use from the twelfth to the fourteenth century, that gave rise to our present-day washbasins. This type comprised a bowl and an 'aquamanile', a sort of copper pitcher in the shape of a fantastic beast or a horseman. In the sixteenth century, the bowl became oval in shape and the pitcher was replaced by a brass or silver ewer. This ewer, which was sometimes engraved and which had a spout and a handle, was the forerunner of the water jug, in a period in which running water did not exist.

The bowl and jug formed the washstands that our grandparents used. When they were first introduced in the sixteenth century, these washstands, which were apparently Venetian in origin, only had three wrought-iron or wooden legs supporting a washbowl. A water jug was kept in the space between the three legs. In the France of the First Empire, mahogany washstands with four thin legs became popular. The most elegant ones were designed by Percier and made by the cabinet maker Jacob Desmalter. They were sometimes real 'wash tables' with an adjustable mirror and seats on the sides; their use was restricted initially to the richest households, but in the course of the nineteenth century they spread to the middle classes. For want of a washbasin, they made do with a simple bowl, and instead of mahogany they were content with pitch pine.

Within a few generations, the washbasin, which had existed in embryonic form in the sixteenth century, was born in the eighteenth century, acquired its definitive shape in accordance with technical norms in the second half of the nineteenth century and became an everyday object for the upper social classes. Nevertheless, some moralists feared that washbasins might be even more conducive to 'auto-eroticism'[36] than baths, the use of which was becoming established only very slowly, even in Paris. The distrust of contact with water, which had its origins in

34-307.
Bidet de toilette céramique blanche, cuvette ovale à gros bords arrondis avec trous d'arrivée d'eau tout autour, trop-plein à l'avant, vidage à bascule, siphon plomb, robinets cuivre nickelé avec index porcelaine "**Froid**" et "**Chaud**" et raccord pour tuyau plomb de 12 × 22ᵐ/ₘ, long. 62ᶜ/ₘ, larg. 38ᶜ/ₘ, haut. 42ᶜ/ₘ, poids 20 kgs. *Livré avec vis tête nickelée pour le fixer au plancher*.............................. **310.** »

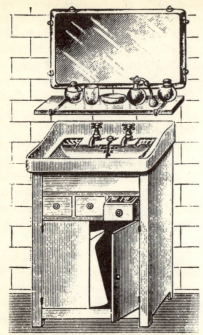

34-312. Lavabo confort pour chambre, cabinet de toilette, composé d'un lavabo nº **34-252 bis**, complet, posé sur un **meuble bois dur laqué blanc,** 3 tiroirs pour brosses, peignes, objets de toilette fragiles ou intimes. Dessous pour seau, broc, serviettes, masquant la vue des canalisations. Dim.: larg. 65ᶜ/ₘ, prof. 45ᶜ/ₘ, haut. 90ᶜ/ₘ Poids 32 kgs env. *Ce lavabo est le plus pratique et le plus confortable entre tous ; entièrement laqué, il n'exige aucun entretien. Recommandé pour chambres, lorsqu'on ne possède pas de cabinet de toilette.*
Prix.................................. **585.** »

1931. Extract from the Manufrance catalogue, advertising 'modern' washbasins and bidets connected to the urban system. The catalogue emphasizes comfort, over hygiene. (Photograph: © D.R.)

deep-rooted popular belief, was based on a symbolic code: since the bath symbolized 'the turning point between life and death',[37] it was barely possible to take a bath more than two or three times in the course of a lifetime: at birth, on the eve of marriage before changing 'state' and shortly before being wrapped in the shroud. Otherwise, as in some villages in ancient France, 'baths were taken in sweat.'[38] As is the case with other present-day sanitary equipment, the washbasin is the result of technical deliberation and know-how dating essentially from the nineteenth century and inspired usually by the British model. These apparently very simple innovations, which are now in everyday use, have to be assessed at their true worth.

Thus washbasins are now supplied with water by means of taps located above the basin. This was not always the case, particularly in Britain, but also in France, especially in the cities, where British models were copied almost unthinkingly. Experience showed that the position of the taps was not only a question of aesthetics, but that it had to be dictated by principles of hygiene. It became common practice to place the taps above the basin in order to prevent them from being constantly deluged by the water used for washing. Similarly, it was discovered that the water had to be drained away directly and sufficiently quickly, that each basin, in order to be clean and efficient, had to have an S-bend and that the overflow pipe had to be plumbed in above the S-bend. Although these design principles appear obvious today, it took time for the sanitary engineers of the last century to discover them.

Baths and bathrooms

Fixed or portable, public or private, the bath goes back at least to Homer's time. Made of wood, marble, sometimes even of silver, it was usually shaped like a tub and bore little resemblance to the bath we know today. The elongated shape of our baths did not emerge until the end of the Roman Empire, under the name of *alveus*.

In the Middle Ages, the bathtub was distinctly more rustic. It was made of wood or metal and was usually circular in shape; it was generally to be found in the public bath-houses that were distant descendants of the public baths of antiquity. Baths were also to be found in private households, although only in those which could afford this supreme luxury which did away with the need to go to the public bath-house.

It was during the reign of Francis I that baths in the modern, elongated shape began to appear. A century later, they were made of copper, tinplated on the inside and with painted decorations on the outside. They were to be found in rich dwellings, often in the bedroom, and covered with a canopy from which hung muslin or linen curtains. Only permanently fixed baths, such as those intended for use in hospitals, and the portable baths brought on request to private houses, together with the necessary buckets of water, were made of wood.

It was between 1770 and 1900 that most of the major technical developments took place. Firstly, the material from which baths were made began to change. From 1770 onwards, baths began to be made of sheet metal, which was less expensive than copper and also more hygienic when coated with a special enamel-like paint. By about 1840, zinc was in use as a material, to be followed in about 1880 by cast iron and around 1900 by porcelain and ceramics.

Secondly, water was no longer heated in cauldrons on the hearth or on the cooking stove, nor even by the immersion of a tank full of hot embers; water-heating devices running on wood and coal appeared from 1840 onwards. By about 1870, the first gas-fired bath heaters, inspired by the English model, entered the market, while the 'modern' mains-fed water heater was being developed.

Finally, the arrival of a supply of running water and connection to the mains drainage system made it much easier to install baths in the home.

On the other hand, in working-class homes, not only at the end of the eighteenth century but for a long time afterwards, 'water was a rarity' and a 'luxury'.[39] It was 'precious and costly'.[40] In these houses, bathtubs were replaced by buckets and tubs, and more particularly by 'baths taken in the river or in public bath-houses',[41] which were more or less expensive and which increased in number throughout the nineteenth century.

Although baths had their origins in antiquity, bathrooms, which were first developed in England, appeared for the first time in France in the 1730s, during the Regency period. In this age of 'debauchery', the glorification of the body, similar to that which prevailed during the Renaissance, seemed to accord perfectly with the desire for comfort, intimacy and discretion.

The bathroom, a place set aside for nakedness and twin sister of the water closet, combined 'luxury, tranquillity and delight' and contributed to the development of specialized rooms, which until that time were virtually unknown. It encouraged the birth of a set of actions and rituals associated with cleanliness and beauty, in which the body was given up for inspection in mirrors and, for some time to come, by the eyes of a maidservant.

Because of their high cost and exoticness, private bathrooms for a long time remained the prerogative of the upper classes. On the eve of the Revolution, the wealthy court surgeon Caignard followed the royal example. He had a princely residence built at Versailles, with a cellar, orangery, winter garden, billiard room. . . and bathroom.[42]

However, this was an exception. For a long time to come, indeed for virtually the whole of the nineteenth century, bathrooms remained a luxury restricted to very few people. One of these was the prefect of the department of Nièvre, who in 1840 was the only person in the department to have a bathroom; another was Zola's Nana, whose bathroom was terribly ostentatious. Nor were bathrooms common in the homes of the Parisian bourgeoisie, even in about 1880.[43] This was perhaps because, in the words of the Comtesse de Pange, 'the idea of immersing oneself up to the neck in water was considered pagan.'[44] It was probably also due to the fact that the use of bathrooms required some sort of central heating system in the home, which did not become widespread until about 1860, and also to the high installation costs. Throughout the whole of this period, the well-to-do classes generally made use of public baths, which were the direct ancestors of our municipal shower baths which are now beginning to disappear. Otherwise, they used the 'domestic bath services' offered in Paris by private companies which supplied both the water and the bath. As for the working classes, they bathed in the river in summer, or else made do with the 'tuppenny baths', 'a few of which, enclosed by crude wooden fences, were set up in Paris directly on the river.'[45]

Between 1817 and 1831, the number of public bath-houses, which totalled almost 500 in 1816, increased rapidly as a result of the water brought to the city by the Canal de l'Ourcq. In 1831, the total was 2,734. In addition, there were 1,059 companies providing a domestic bath service and 335 providing baths on boats, to say nothing of the hospitals and poorhouses which provided baths for the public.

If baths are counted rather than bathtubs, the number of baths taken in Paris in 1852 was almost 2,778,400. This total can be broken down in the following way:

1,818,500 baths in bath-houses;
   350,000 hot baths in establishments located on the river;
   109,900 baths in hospitals and poorhouses providing a public bath service;
   500,000 baths in the 21 cold bath establishments set up on the Seine and the Canal Saint-Martin.

The French law of 3 February 1851 on public baths and wash-houses had to a certain extent encouraged this increase by granting credit of 600,000 francs in order to subsidize, up to a maximum of one third of the cost, those local authorities which requested the creation of establishments of this type; this made it possible to provide baths free or at low cost. In about 1850 in Paris, luxury baths cost between 5 and 20 francs, while the daily wage of a manual worker might be 2.50 francs. In contrast, baths for working people cost 1 franc, falling to 0.75 francs in about 1860. The average price was about 0.60 francs and even as low as 0.45 francs for subscribers.[46]

Bathrooms, which were still a recent innovation, did not become a widespread phenomenon in towns until relatively late, unlike thermal baths, bath-houses and shower baths. In 1931, 57 per cent of Italians had a supply of drinking water, but only 10 per cent had a bathroom. In 1954, only 1 dwelling in 10 in France had a bath or shower.[47] Even today, not every house has one. In 1968, barely 17 per cent of French agricultural workers had a bathroom in their homes.[48]

And yet as early as 1889, Jules Verne had in his own way anticipated the future proliferation in his *Day in the life of an American journalist in 2889:*

'There is always a bath prepared in the hotel, and I do not even have to bother to leave my room in order to take it. If I just press this switch the bath will start moving, and you will see it appear all by itself with water at a temperature of 37° C.' Francis pressed the switch. There was a muffled noise which swelled in volume. . . . Then one of the doors opened and the bath appeared, sliding on its rails. . . .

The bidet

The trilogy of which the washbasin and bath constitute the first two elements would be incomplete without the bidet which came into widespread use in the nineteenth century, but which has a long history. It was first used in France in the sixteenth century, and it is assumed that it was imported from Italy during the reign of Francis I.[49] The Musée de Sèvres today has several examples of this oblong receptacle designed for washing the genital and anal area.

This 'porcelain violin case, mounted on four legs' (to quote a prudish nineteenth-century auctioneer),[50] was not in common use among the upper classes until the beginning of the eighteenth century. The first mention of it in literature would seem to be in the *Journal* of the Marquis d'Argenson, who describes the attempts to seduce him made by Jeanne Prie, an extremely dissolute woman who was at the time the mistress of the Duc de Bourbon and who was thought to be afflicted by venereal disease.[51]

During this period, the greatest ladies in France had a bidet installed in their water closets. The Princess of Savoy who had just married the Count of Provence, received 'a red wooden bidet with a bowl of silverplated copper in a leather case'; moreover, it stood next to 'a close-stool [. . .], with a crimson velvet cushion and its own bucket in silverplated copper in a leather case'.[52] The same wedding present was given during the same period to the young Marie-Antoinette when she arrived in Strasbourg on her way to marry the Dauphin. She was sent 'a bedside table, a bucket to wash her feet in, a fully-fitted bidet and a close-stool'.[53]

Madame du Barry had a silver bidet with a sponge box. For Madame de Pompadour, Pierre de Migeon, the creator of the knee-hole writing table, made a splendid walnut bidet 'with a lid and back of red leather with goldplated nails and two crystal flasks'.

In 1739, Rémy Pèverie, a cabinet-maker in the Rue aux Ours in Paris, designed double back-to-back bidets which could be used by two people at a time. In 1762, Jean-Baptiste Dulin, a fellow cabinet-maker, improved the furniture available for use on journeys or in the country by designing metal bidets fitted with feet that screwed on or off. And in the years just before the Revolution, Jean-Baptiste Cochois, who worked in the rue Saint-Honoré, specialized in making furniture that concealed its true purpose: he made a chiffonier that converted into a bedside table, commode and bidet.

In the eighteenth century, the bowls were usually made of earthenware or porcelain: Nevers earthenware with polychrome floral decoration, Rouen earthenware with a blue background, Moutier earthenware decorated with grotesque figures drawn in green, or India

Company porcelain with a floral decoration and the French coat of arms. Mounted on legs made of valuable wood, bidets were often fitted with a backrest and sometimes even equipped with syringes for use as douches.

Frightened no doubt by the infernal heat of sex, the nineteenth century shrouded the bidet in silence until the era of Napoleon III. Only prostitutes, women of easy virtue or courtesans were said to use them for contraceptive purposes which were adjudged scandalous. If the hygienist Parent-Duchatelet is to be believed, cleanliness among prostitutes improved considerably between 1811 and 1836, 'as a result of the vigilant action of doctors and inspectors of brothels'. 'They have at least acquired the habit of washing, to an extent that might even be called excessive, and which is only neglected by those among them who are brutalized and degraded.'[54]

During the second half of the nineteenth century, it was in brothel rooms that the bidet occupied a place of honour. At the 'Chabanais', one of the most famous in Europe — a veritable annex of the courts of appeal and the Jockey Club — the Parisian proprietor, worried about his income, had a special filter installed on the bidets used by his ladies, in order that the hard water should not damage certain tissues considered too delicate by far. By about 1880, the use of the bidet among prostitutes was reflected in slang: a pimp was known as a 'knight of the bidet' and he was said to 'obtain his bread from the bidet'.[55]

Whether from prudishness, discretion, hypocrisy or mere disinterest, the bidet did not figure among the major concerns of the great hygienists. It is not included in the main nineteenth-century dictionaries of medicine and hygiene, at least not under that name. It was not until the *belle époque* that the *Nouvelle Encyclopédie pratique du bâtiment et de l'habitation*[56] and a book written by the engineer Marcel Hegelbacher[57] dealt with the subject and recommended that bidets should be supplied with hot and cold water. Outside France, there is little evidence of the existence of bidets in nineteenth-century Europe.

From 1920 onwards, bidets came into widespread use in French towns. They were made of 'sanitary porcelain' imported from the USA, which was impermeable and resistant to acids and bases. They featured in all the catalogues of the major French manufacturers and advertisements for them multiplied, for example in *L'Illustration*.

The origins of the bidet are uncertain, but it is French by reputation; in the 1780s, it called forth the surprised admiration of the Englishman Arthur Young:

The notion of cleanliness is different in the two countries. The French pay more attention to personal cleanliness, while English houses tend to be cleaner; I am speaking now of the mass of the people and not of the very rich. Every apartment in Paris has a bidet, as well as a bowl for washing the hands. This is

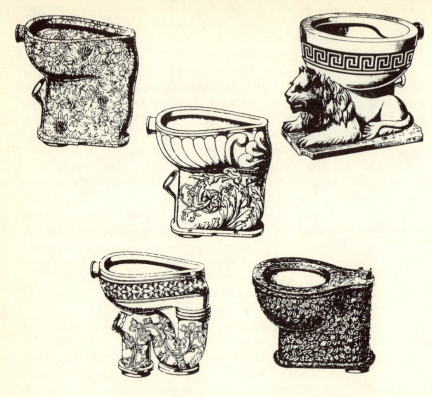

'Les fleurs du mal' (1890–1895). The ostentatious decoration conceals the embarassing aspects of these lavatory bowls, which were connected to the main drainage system.

an aspect of personal cleanliness which I would like to see become more widespread in England.[58]

Nevertheless, examination of post-obit valuations in Paris shows that, in 1850 and 1880, less than 5 per cent of the inhabitants of the capital had a bidet.[59] Around 1900, according to the Goncourt brothers 'there is not a single bidet' in Saint-Denis, Ecouen or Picpus, while in Paris 'the devout do not wash their bottoms.'[60]

Accused by moralists of encouraging oral sex, it took a long time for the use of bidets to gain widespread acceptance. At the end of the nineteenth century, the Ducs de Broglie, who had had water and water closets installed in their manor house in Anjou, refused to have a bathroom which was a 'useless luxury found in large hotels and which in private houses might even be called blameworthy'.[61] These prudish traditions largely explain the slowness with which the bidet was introduced into French homes.[62]

The 'seat of vanity' first appeared in France at the beginning of the eighteenth century. Iconography suggests that it was used exclusively by women. (*Woman at her toilet*, by Boilly.)
(Photograph: © Roger-Viollet.)

Water closets

It was at the beginning and, more especially, at the end of the nineteenth century that a certain number of new practices in the area of the so-called 'natural functions' were developed.

For centuries, in Paris as elsewhere, it had been the practice simply to throw all waste and rubbish into the streets, or into the Seine. There were, however, a few exceptions: between the twelfth and fourteenth centuries, the chateaux of Coucy, Chauvigny, Marcoussis and Pierrefonds had latrines connected to drains, or towers used as lavatories. In Paris in the eleventh century cesspits were installed outside dwellings.[63] In 1387, Queen Isabeau of Bavaria purchased a chamber pot virtually identical to those used by our grandparents. Her paymaster gave 32 sous

1910. Extract from the catalogue of a Parisian department store. Portable bidets on stands, without a water supply. They were usually kept out of sight.
(Photograph: © D.R.)

minted in Paris for 'a double case of *cuir-bouilli* for depositing and transporting the Queen's urine, stamped and emblazoned with the Queen's coat of arms, and with a key for locking it up'.[65] The height of refinement, however, was the silver chamber pot that Louis XIV kept under his bed and which was stolen from him.[65]

The receptacle was in fact of little importance. The custom was to throw urine and excrement out of the window, as is shown by the royal interdicts issued from the sixteenth to the eighteenth centuries.[66] Moreover, it was a long time before every Parisian householder installed a cesspit, despite the decree of 26 August 1531. It is true that cesspits were not installed inside Parisian houses until this same period.[67] However, a long time after the sixteenth century, the cry of 'look out below!' meant that one had to take to one's heels: on 5 February 1682, in Moustiers in Provence, this practice gave rise to an incident. Antoine Langier

having sat down on the stone outside the house of Jacques Fabre, called out to Isabeau Petit, the wife of Jean Féraud, saying that 'he was troubled by haemorrhoids and that she should put out her light, since he wanted to go to the toilet'; however, while he 'was on the said stone', Isabeau Petit took hold of a chamber pot 'containing filth, into which she put water' and threw the entire contents over his person! When questioned, she stated 'that it might indeed be true that she threw water from the window that evening, but it was clean water that she threw away every evening, not with the intention of harming anybody, but simply in order to get rid of it, since she had nowhere else to throw it because the rear of her house adjoined the church'; finally, when questioned if she was familiar with the regulation prohibiting the throwing of filth, water or anything else out of the window 'without shouting a warning three times', she replied that she could not remember whether or not she had taken that precaution.[68]

And neither Molière[69] nor Scarron missed an opportunity to poke fun at the bashful lovers who, mooning under the balcony of their beloved, found themselves covered in a foul-smelling stream of water. Thus Don Japhet, at the cry of 'look out below!' uttered by a vigilant chaperone, cries out in his turn:

> Gare l'eau! Bon Dieu, la pourriture.
> Ce dernier accident ne promet rien de bon.
> Ha! chienne de duègne, ou servante ou démon,
> Tu m'as tout compissé, pisseuse abominable,
> Sépulchre d'os vivants, habitacle du diable,
> Gouvernante d'enfer, épouvantail plâtré,
> Dents et cris empruntés, et face de châtré[77]

(Look out below! Good God, you swine. / This last mishap smells quite evil. / Ha! cursed chaperone, whether maid or demon, / you've drenched me in piss, abominable woman, / You bony sepulchre, wherein dwells the devil, / Diabolic governess, painted scarecrow, / With borrowed teeth and cries, and weakling's face.)

In fact, even in towns, the privies[71] installed on the ground floor of certain houses were hardly ever used. Close-stools, which were at the height of their popularity in the seventeenth and eighteenth centuries, remained very much in use. However, the way in which they were used did change in one important respect: by the end of the eighteenth century they were no longer used publicly but rather in private.

During the reign of Louis XIV, they were still in use in the best circles. They were a highly valued stronghold on which one could daydream for a while. They could be used for meditating, dreaming, chatting, writing and playing, by generals and ministers as well as by society ladies.

At the beginning of the eighteenth century, Louis-Joseph, Duc de Vendôme and the great-grandson of Henri IV

used to rise fairly late in the army and would sit on his close-stool, attending to his correspondence and giving his orders for the morning [. . .]. This was the time when those who had business with him, that is officers and important people, could speak to him. [. . .] He ate a hearty breakfast there, with 2 or 3 intimate friends, and passed an equal quantity, either while eating, listening or giving his orders [. . .]. He passed a great deal; when the bowl was full to overflowing, it was taken out and passed under the noses of the assembled company in order to be emptied, often more than once [. . .]. He claimed that the simplicity of his habits was worthy of the early Romans and put to shame the ostentation and extravagance of others [. . .]. The Duke of Parma had to negotiate with M. de Vendôme; he sent as his representative the Bishop of Parma, who was very surprised to be received by M. de Vendôme on his close-stool and even more so to see him get up in the middle of the audience and wipe his bottom in front of him. He was so scandalized that, without saying a word, he returned to Parma without concluding the business that brought him there, and declared to his master that he would never in his life return after what had happened to him.[72]

The habits of the Princess Palatine were no different. As soon as she rose, she would sit for a long time on her close-stool,[73] and, full of wit in this respect also, tells us that the famous shares issued by Law as part of the Mississippi Scheme for the development of Louisania, which collapsed due to excessive speculation, found their ultimate purpose there: 'Nobody in France has a penny to their name any longer, but with permission and in good palace language, we have bog-paper in profusion.'[74]

Madame du Barry had great affection for it. Jean-Jacques Rousseau daydreamed on it 'for hours at a time'.[75] The Duc d'Orléans received the Duc de Noailles while sitting on it, surrounded by servants.[76] Voltaire wrote to his agent on it: 'My backside, jealous of the beauty of my furniture, requests a pretty close-stool with large replacement buckets. You will tell me that my backside is most insolent to address you in this way, but please remember that this backside belongs to your friend.'[77]

In the course of the nineteenth century, it became common practice to carry out in private those functions which, until that time, had been considered natural and were therefore performed in public. This change was already beginning to take place at the end of the *ancien régime*. In 1738, the architect J.-F. Blondel called the lavatories installed in certain chateaux 'valve closets' or 'English lavatories'. During the reign of Louis XIV, for fear of epidemics and out of disgust at the smell, the director of the royal buildings had public latrines, for which a charge was made, installed on the terraces of the Tuileries gardens. Until then, this role had been played by the groves and bushes, which were of course free — and disreputable. During the Revolution, the Duc d'Orléans had twelve lavatories built in the Palais-Royal; they were 'clean and odour-free' and a charge was also made. Some time later, they brought in an income of 12,000 livres per year.

As a rule, however, '[French] lavatories are temples of abomination.'[78] And Louis-Sébastien Mercier went even further:

Three-quarters of the lavatories in Paris are dirty, horrible and disgusting; in this respect, the eyes and noses of Parisians are accustomed to filth and stench. Hindered by the narrowness of the sites on which houses are built, architects have laid their pipes at random, and nothing can be more surprising to foreigners than the sight of an amphitheatre of latrines perched one on top of the other, adjacent to staircases, beside doors, close to kitchens, all giving off the most fetid odours. The undersized pipes easily become blocked; they are not unblocked, and the excrement piles up in columns, getting closer and closer to the lavatory seat; the overloaded pipe bursts, the house is inundated, infection spreads, but nobody leaves the house: their noses are hardened to these pestilential setbacks.[79]

Between the end of the eighteenth century and the 1840s, certain improvements were made to new cesspits installed, for example, in Paris: 'since cesspits are now constructed with extreme care, our houses now have only impermeable tanks, which retain everything put into them.'[80] Nevertheless 'only fifty years ago [about 1790], the so-called English lavatories were still a luxury that were found only in the richest houses. They are now in almost universal use in Paris; there are no newly constructed or renovated appartments which do not have one and sometimes several such lavatories.'[81] In other words, these lavatories, which were watertight and had a supply of running water, were one of the great innovations of the early nineteenth century, and were made possible partly by the extension of the water mains supplied by the Canal de l'Ourcq.

Before the development of the WC used today, experiments were carried out over a period of several decades with devices and lavatories intended to deal satisfactorily with human waste. A succession of different types of devices and lavatories were tried, some of them even coexisting for short periods; these included dry latrines consisting of a toilet located above the cesspit and lavatories with a cesspit, cask or tank that had to be cleared at night by the cesspit clearers. A variety of different technical solutions introduced water into the new lavatories, which were known as 'English lavatories' or 'water closets'. They were all intended to achieve the same ends: to take the excreta out of sight and to suppress the stench. Embarrassment and prudishness, which had hitherto been absent from an agrarian culture in which excrement was valued for its power as a fertilizer and healing agent, were now fostered by the puritan ideology of the nineteenth century. Within the space of two generations (1860–1914), lavatories became an integral part of the domestic universe. Toilets, or water closets, duly enclosed in the same way as rooms intended for domestic use, were relocated from the courtyard (or garden) to a location alongside the house and then into the building itself, near the routes of communication: entrances, staircases and lobbies. It was no longer permitted to defecate anywhere but in these closets, and urinating or defecating anywhere but in the house, outside in the streets or squares, was socially proscribed.

It is not so long since Louis XIV used to grant audiences while sitting on his close-stool. In 1770, Louis XV had water closets installed at Versailles for his own personal use; they were fitted with a porcelain basin with two water jets, one for emptying and one for cleaning.[82] In the space of a few decades, the appearance of the lavatory changed, as if the raising of the threshold of modesty made people want these very intimate places to take on a certain solemnity.[83] During the Restoration, it was decorated with friezes, Greek borders, acanthus leaves or arabesques and took on the aristocratic appearance of the lavatories of antiquity.

Throughout the nineteenth century, lavatories were in the news. In fact, the most widely recognized technical standards were hardly implemented at all, as was noted by the Commission on Insanitary Dwellings in Paris during the Second Empire, and then during the Third Republic.[84] The watertightness of the cesspits, the system of sealing the lavatory pan, the dimensions of the room, and lighting and ventilation, all left much to be desired.

Such were the genuine anxieties caused by the stench from the cesspits that the lack of hygiene evident in their construction was the subject of many reports,[85] learned treatises,[86] articles in scholarly journals, and entries in highly technical dictionaries.[87]. The anxieties engendered by exaggerated modesty were compounded by the fear of cholera; this was well before the science of microbiology had revealed the existence of germs in faeces! Spread by the notions of 'airism' and contagionism, the fear of malodorous gases was not dissipated by the invention and application of ingenious, if inefficient ventilation devices.[88] In fact, the only efficient procedure was an improved model of the water closet with a hydraulic siphon to ensure that the lavatory pans were completely sealed off. The water closets also had to be well maintained and supplied with water.

For a long period, from about 1850 to 1900, French sanitary engineers and health officers debated the shape, material, advantages and disadvantages of a considerable number of models of WCs, mostly of British or US origin. After much hesitation, it was decided to opt for the British 'pan-closet' model.[89] It was sealed off with a concave valve and the base of the device was formed by a broad pyramid connected at the side to a large box containing the operating mechanism. The technical improvements made during this period were all intended to prevent the risk of contagion; thus the valve had to be so designed as to prevent the entry of 'foul gases' each time it was operated, while the vent had to be stopped from becoming an inlet pipe 'for infectious gases and germs'. Similarly, there had to be a sufficient head of water to flush the waste away effectively. The flushing mechanism had to be different from the system in widespread use in Paris in about 1880, which depended on a tank that had to be topped up manually and which provided an inadequate flow.[90]

On the other hand, the sanitary engineers of the period were of the opinion that this water supply should not be connected to the rising main of clean water because of the risk of back flow. This type of water supply, which had been banned in Britain and the United States for a long time, was not yet prohibited in France in 1884, although French engineers were also aware that 'this practice [. . .] may give rise to the most serious epidemics.' Finally, the WC had to satisfy two principles laid down by the engineer Daniel Emptage in 1883:[91]

> the water in the U-bend had to be visible in the pan and completely protected from the siphoning mechanism;
> the device had to be thoroughly cleansed by the water used to flush the pan and the U-bend of all waste matter.

The modern, individual WC which, in the West, has today superseded all other devices[92] for getting rid of human excrement, is the direct descendant of the British 'water closet'. It was in this way that water was introduced on a massive scale into sanitary equipment designed for purposes of bodily hygiene; it began to play an increasingly important role and its ancient symbolism acquired fresh vitality as a result. Within a few generations, between 1830 and 1840, it had become the established practice to wash and attend to one's bodily functions in private. The result was the development of real control over one's body. Dirtiness, 'noxious odours' and excrement were banished.

The objects and places associated with water that were introduced into the domestic world (filters, washbasins, bidets, WCs) led gradually to the standardization of behaviour related to the body. The concern with hygiene, cleanliness, even purity, brought long-established habits into the purely private sphere: the thirst for pure water, the comforts provided by an abundant supply of water, the evacuation of dirty water contaminated by excrement and 'hygienic' washing all helped to further the new water code and to ensure its conquest of society.

Household filters

Although they were not in widespread use in the 'Old World' around 1760, the use of household filters in France began to increase from 1840 onwards.[93] Even around 1900, their usefulness was not questioned. In cities endowed with a water supply network, 'household filters are often used to supplement the central filtration process; they are all the more useful in areas in which the water supply, although not spring water, reaches the town in more or less the same state as it leaves the watercourse.'[94]

There was a wide variety of different domestic filtering devices, but they could be divided into household filters and table filters; these latter were small and portable enough to be taken on journeys.[95] As was the case with central filtration plants, two types of filter were used: clarifying filters, made of porous sandstone, layers of sand and gravel and a porous mass of wool that had been scoured by boiling, and purifying filters, which exploited the properties of charcoal.

At the Home Economics Exhibition of 1872, two models of carbon filter were shown, together with Petitpas's filters. These latter, which were fitted with a device to facilitate cleaning, prevented the build-up of muddy deposits that imparted a bad taste to the water and made it possible to combine the two processes known at that time: mechanical clarification and chemical purification.

The Piebourg charcoal filter also combined these two methods of 'purification'. However, the superimposed layers of sand and charcoal were supplemented by a sponge which prevented the passage of grains of sand. This was a filter that was particularly well suited to untreated river water full of silt and sand.

Household and table filters that use the so-called plastic (porous) charcoal of C. Buhring consist of a block of porous charcoal immersed in water and fitted with a siphon, the longer arm of which conveys the filtered water into the container. It is possible, if desired, to make the water even safer by filtering it twice. In addition to these household filters, which function with one charcoal block and two buckets, the inventor has devised a table filter consisting of two glass containers placed one above the other; the upper one, which contains a charcoal siphon, is half-filled with water; the user sucks on the siphon, places the longer arm into the lower container, fills up the first container and the filtration process then proceeds of its own accord.[96]

At the 1884 Hygiene Exhibition held in London, two other recently invented filters attracted the attention of specialists.

The first of these is Maignen's 'rapid filter' [which], irrespective of its mechanical effect, acts chemically in a very vigorous manner; it partly burns alcohol, retains lead, iron and some of the salts of these metals and oxidises organic materials.

The second is the Chamberland filter, designed by this scientist in accordance with the filtration principle, [. . .], which retains bacteria and vibrios. A sort of porcelain test tube is fixed by its free edge, closed end uppermost, to the bottom of a metal cylinder 5 centimetres in diameter and 25 to 30 centimetres in height, which covers and contains it by means of a tight hermetic seal. Water at a pressure of between one and four atmospheres is piped into the space between the two tubes; it can only escape by passing through the walls of the porcelain tube, whence it is drained off through a simple tap fixed to a socket on the metallic outer casing. In order to clean the filter, the lower container is unscrewed and the porcelain is washed with boiling water or even heated red-hot by a gas burner. The water that comes out of this filter is a sterilised liquid.[97]

Of course, the proper functioning of any of these household filters was heavily dependent on water pressure, the initial degree of impurity and adequate monitoring and cleaning of the filter. In the case of the Buhring filters, Dr Fonssagrives strongly recommended in 1876 that 'the charcoal should be cleaned after it has been in use for some time [?]; the user should place it in clean water, blow strongly through the siphon and clean and brush the porous charcoal. It should then be left to dry in the sun and a scrubbing brush or file used to remove the impurities on the surface.'[98]

The discoveries of microbiology, one of the origins of which is to be found in the experiments conducted by Pasteur in 1863 and 1864, were to deliver a mortal blow to these domestic devices that now gladden the hearts of second-hand dealers. By about 1900, ozone was being used to purify water on a large scale, and at more or less the same time the water carriers disappeared from Paris.[99]

Washing

Today, washing machines are no longer considered a luxury but as one of the first 'investments' in household appliances. The installation of domestic water and energy supplies made it possible to wash and launder clothes in the home. Washing machines have only been in widespread use in France for a quarter of a century, just slightly ahead of dishwashers . . . which await their own historian. The practice of washing has now been reduced to its functional aspect, without there being any fixed or prohibited days, as used to be the case with Fridays, the day of Christ's death, when it was feared that washing sheets would be tantamount to washing one's shroud.

Few people realize that washing and laundering are two distinct operations. To launder clothes is to return used and soiled linen to its pristine whiteness. It involves the use of a detergent and the removal of dirt and germs by mechanical and chemical means. The term washing, which is increasingly replacing the term laundering, implies a more gentle process which can be carried out with soap alone, or a mild detergent. Finally, in contrast to WCs, which are places set aside for the production of ordure, the area set aside for washing is uncertain; washing and laundering is done in the bathroom, the kitchen or the basement, anywhere where there is a supply of water and energy and a connection to the drainage system.

There is a considerable gap between the practices of times past, that is before 1900, and those of today. The changes that have taken place are due as much to social and cultural changes as to the technical developments that have affected domestic laundering.

From the end of the eighteenth century, in Paris,[100] the task of laundering was taken out of the home. An elite group of scientists – notably chemists such as Parmentier (1737–1813) and Chaptal (1756–1832) – in conjunction with industrialists, property owners and economists, overturned traditional procedures and transformed an ancestral practice into an applied science. In 1805, Cadet-de-Vaux (1743–1828) published a 'Popular Instruction Manual on Steam Laundering', followed by a 'Treatise on Steam Laundering'. The washing was soaked by a process of 'affusion' in steam, and industrial laundries and those in large organizations such as hotels and hospitals were used to test the economic and mechanical performance of the new system. The revolution in chemistry introduced chlorine, and bleach transformed the process of laundering from 1784 onwards; a few minutes now sufficed for this operation, instead of several weeks. Common soda (or a sodium salt), used with cinders, formed the detergent. From 1790 onwards, it was replaced by much cheaper synthetic soda crystals, manufactured by the Leblanc process, which was superseded from 1865 onwards by the Solvay process.[101]

While the traditional washing process persisted for a long time in small towns and the country, modern practices became increasingly more common in country houses and middle-class homes and in industrial laundering and large institutions such as hospitals, barracks and boarding schools. In country houses and private town houses, laundry rooms were installed and equipped with washing machines, clothes driers, buckets and wooden mangles. Here servants – mainly women – washed the clothes of the well-to-do; otherwise, the aristocracy and the middle classes sent their washing to washerwomen or to industrial laundries. In imitation of Saint-Simon, a few well-meaning philanthropists, such as Harel in 1839, provided the tenants of his housing association in Paris with houses with central heating, bathrooms and double walls. A modern community laundry was set up, which reduced the cost of washing clothes. The whiteness of the clean linen was shown off to advantage at the communal table. Slowly, the places, devices and practices characteristic of modern washing spread from the top to the bottom of the social pyramid. In short, gracious living came to city dwellers.

On the other hand, in peasants' hovels, in workers' dwellings and in the 'dripping courtyards' of Lille[102] there was rarely a supply of water and water for washing had to be fetched from the public fountain or the nearest well.

Otherwise, just as in the eighteenth century, one had to go to the river to wash one's clothes, even if one was a man! Indeed,

A man who has only one or two shirts does not surrender them to the washerwoman's paddle, but plays the washerwoman himself in order to preserve his shirt. And if you do not believe this, go to the Pont-Neuf one Sunday in summer, at 4 o'clock in the morning, and you will see several individuals on the river bank, dressed in what appears to be a frock coat, washing their only shirt or their one handkerchief. They then hang this shirt out on a miserable stick and wait for it to dry in the sun before putting it on.[103]

Although there was no shortage of philanthropic intentions in the nineteenth century, widespread change was, with a few exceptions, slow in coming. It was not until the end of the century that modern washing methods spread to the home. A whole range of legislation contributed to the spread of these new practices: the regulations on the combined sewer system in Paris (1894), the law on low-priced housing (1894) and the law on communal sanitation (1902), which obliged mayors to enact sanitation bylaws. The period following the First World War saw the rapid spread of urbanization, and dwellings were no longer designed solely in accordance with the norms of public health. In 1931, Le Corbusier stated that 'domestic appliances [. . .] will become [. . .] a veritable source of happiness.' Since the 1900s, the little domestic boiler made of galvanized sheet metal had entered many homes; it heated water on wood or coal stoves, on cookers or on gas rings.

As early as 1875–6, bourgeois homes in Nevers had them, and they made much appreciated presents for newlyweds. This technical innovation had some important consequences: it was now possible to do the washing once a week rather than two, three or four times a year; it was also possible to sort through the washing, and dirty linen no longer had to be stored up. Centres of infection were suppressed; finally, saponification – and thus hygiene standards – proved to be much improved by the use of soda crystals and detergent. Even though some housewives complained that it wore out clothes more quickly, the boiler meant that one's wardrobe could be smaller and less costly. However, its popularity grew only slowly after 1914, and the wash tub and the slow chemistry of the embers remained in favour among older people until about 1940. Finally, although the boiler made women's work easier by changing the soaking process, it did nothing to change the processes of soaping and rinsing, which were both long and laborious operations.

Research on washing machines was being carried out in about 1840, but the first mechanical washing machines (handle-operated) did not appear until the beginning of the twentieth century and were sold in small numbers only after 1914. Automatic washing machines (with electric motors) began to come into widespread use between 1950 and 1960, thus signalling the end of long-established traditions. Although they did not banish the old vapour, they did lead to the demise of washerwomen who charged for their labours by the item or the day, and in Nevers as throughout the West 'one now hardly sees women rinsing washing in their tubs on the banks of the Loire.'[104]

Tasks and functions that had once been performed in public, particularly the washing of clothes and defecation, were now relegated to a distant, even somewhat barbaric (sic) past. The layout of dwellings was transformed. Rooms were adapted for specialized purposes and many customs and practices were brought into the private sphere. 'Modern comforts' were installed, and with them came a whole new art of living. The urban landscape, both on the surface and underground, also underwent change. Such had been the evolution of the thresholds of perception (sight and smell) that 'a physical and moral geography of cleanliness'[105] ousted the old landscape. Glorification of the fecund qualities of excrement gave way to a feeling of shame. The naked body was hidden from sight. The canons of hygiene were intended to replace the old conviviality and codes of civility, which were now considered rather stupid, even obsolete. The 'house of death' described by Karl Marx in 1884[106] gave way to the 'house of light' so beloved of Le Corbusier and, long before him, of the founder of the workers' cooperative associations, 'where there is ample provision of air and water'.[107]

3

The triumph of hygiene

Despite the uncertainties that surrounded the medical sciences, or perhaps because of them, hygiene became institutionalized at the beginning of the nineteenth century. Even concepts that we *now* know to be inaccurate had some merits. Together with the epidemics that wreaked havoc on both sides of the Atlantic from the 1830s onwards, they gave rise to much deliberation and inspired municipal authorities to install sanitation systems and to introduce other town planning measures. Topographical and social analysis of the spread of these epidemics revealed the links between poverty, overcrowding and lack of sanitation; in London and Leeds, Paris and Brussels, the great predecessors of the socio-demographic survey provided ample demonstration of these links. Among those who made great contributions to these surveys were the Frenchmen J.B. Parent-Duchâtelet (1790–1836) and L.R. Villermé (1782–1863), the Belgian L.A. Quételet (1796–1874) and the Boards of Health of various British cities.

Analysis of this kind was the main basis for the great town planning projects to which many of the cities of Western Europe (and of the East Coast of the United States) owe their present appearance. In the name of progress, the circulation of air was encouraged; miasmas were banished; light was increased and the benefits of sunlight were made available to the greatest possible number of people; water supply and drainage systems were improved, and dwellings were made more sanitary. It is true that these large-scale programmes, such as those launched in Vienna, Berlin, Baltimore, Paris and Bordeaux, mainly involved the city centres and were not inspired solely by notions of hygiene and sanitation. National or regional prestige, social segregation and the desire to thwart certain revolutionary endeavours also played their part. Nevertheless, even though such projects did not involve working-class districts and suburbs until much later, the crusade was under way.

Although there were marked variations from one country to another, it involved all the Western countries, on both sides of the Atlantic. In the Germanic countries the tradition was long-established and went back to the *Aufklärung* and the second half of the eighteenth century. Johann-Peter Frank, founder of the *Medizinische Polizei*, had been its enlightened propagandist. In the Anglo-Saxon countries, 'sanitary awareness' developed at the same time as the notion, if not the current expression, of social policy. In France, the terms 'polity' and 'health' were somewhat at variance with each other. Certain people refused to accept that a well-ordered society was a mark of civilization and that the statutory regulation of public health was one step on the road to an orderly society. Nevertheless, under one label or another, modern notions of hygiene triumphed shortly after 1800, and the phenomenon gained strength with the passage of time.

The case of France

During the first half of the nineteenth century, France set the international pace in matters of hygiene.[1] The French movement had its roots in the philosophers of the Enlightment and their ideological successors, the legislators of the Consulate and the First Empire. Hygiene was included in the original curriculum of the state schools of health [Ecoles de Santé] set up in 1795 during the First Republic; and the first holder of the chair of hygiene in Paris was the illustrious Jean-Noël Hallé (1754–1822). In 1802, the Council for the Salubrity of the Seine was founded; its members included Parmentier, Thouret and Cadet de Grassicourt, who were more than just famous. Other health councils were subsequently established: Marseilles (1825); Lille and Nantes (1828); Troyes (1830); Rouen, Bordeaux, Toulouse and Versailles (1831).

In France, as in the other major European countries, there were periodicals devoted to hygiene,[2] some of which have been preserved for posterity. These include the *Annales d'hygiène public et de Médecine légale*, founded in 1829, and the *Annales des Ponts et Chaussées*, which dates from 1831.

It subsequently became an important subject in the faculties of medicine, where chairs of hygiene were set up. Since the Enlightenment, hygiene had been presented in Parisian and provincial academies in the form of observations, theses and competitions; it now became identified with a Parisian institution which, in the second half of the century, had national influence. This was the Consultative Committee on Public Health, which subsequently became the Higher Council for Public Health. From 1848 onwards, moreover, a Health Council exercised powers, theoretically, in each department in France. At symposiums and conferences, technical standards were widely debated, particularly in so far as they impinged upon the social, demographic and health policies of governments. Finally, hygiene was the subject of a whole arsenal of legislation and statutory regulations, as a result of which it gradually became incorporated into a new medical framework which claimed to be based solely on scientific premises.

The case of Germany

Until 1866, Germany was a piecemeal collection of states that Heinrich Heine described as politically 'fragmented'. It was made up of powerful kingdoms, small duchies and free city states with differing political regimes. The first loyalty of the inhabitants of these various states was to their own state rather than to Germany as a whole, even though the German language, the Lutheran tradition and the *Zollverein*, established in 1834, strengthened the ties to the detriment of the divisions.

As is well known, two conflicts enabled Prussia to dominate the other German states. Shortly after the battle of Sadowa, Bismarck's attempt to unify Germany had its first success. In 1866, the North German Federation was set up, under the leadership of Prussia.

Whether by happy accident or disturbing coincidence, it was in the new German state, now stronger and more closely united, that in 1867 the Society of German Naturalists and Doctors established a new department devoted exclusively to matters of hygiene. In 1869, this department published the first specialist German periodical in this field. It was then opened up to municipal officials and became independent in 1873 under the name of the Society of German Hygienists. Although there was still a small majority of doctors, its members also included professors of letters, magistrates, jurists, engineers and industrialists. From the beginning, the Society supported Bismarck's policy in the hope that the state would direct and centralize public health policy. However, the founding father of German hygiene was not involved. Virchow, the promoter of social hygiene in Germany and, moreover, the co-founder of the Socialist Party, was always to remain on the margins. On the other hand, the Bavarian Pettenkoffer (1871), whose authority in matters of hygiene was so controversial, was a member of the Society.

From its first campaign, the Society enjoyed a very high reputation (1869). These hygienists recommended the creation of a central body which would promote, organize and supervise public hygiene within the confederation. This was tantamount to becoming established as a natural ally of the leader of the Confederation (the kingdom of Prussia) and to setting up as the mouthpiece of all German hygienists. And yet the situation was not so bad in Saxony, Prussia or Hesse, where public health had been organized for several decades. The ambitions of the Society and its periodical were considerable and, in 1869, virtually unacknowledgeable. The aim was to equip all the German states with the same legislative and administrative framework as the Prussian model. There is no doubt that, as in France, the British example played a role. British legislation in the area of public health was distributed and analysed, such was the fame of the British Public Health Reports and the power of a society of which four of the five founder members were Berliners.

In fact the centralization championed by the Society raised problems. When the Second Reich came into being in 1871, members of parliament were divided on the role of the state; the desire to preserve autonomy and particularisms was still very strong. Finally, after several mishaps, the *Reichsgesundheitsamt*, the Imperial Hygiene Department, was set up in 1875. Nationalist, tinged with Anglophilia and hostile or indifferent to the French example, the Second Reich set up a central body whose function it was to collect statistics and to implement the most general hygiene measures (social welfare, organization of the medical professions, prevention of infectious diseases). However, each German state retained its independence in the organization of public health, sanitation and sanitary regulations.

The power devolved upon institutions in the field of hygiene depended on their dominant characteristic. The power of observing and advising – the expert role – and of decision making – the 'standardizing' role – was the prerogative of the scientific institutions. The partly scientific, partly administrative institutions, like the departmental Health Councils, were given the task of sending observations to higher bodies and of implementing legislation and statutory regulations relating to public health. Political institutions, such as the National Assembly and Senate in France, were given the power of legislating, notably in 1884, to increase the powers of the municipalities and, in 1902, to introduce a vast, wide-ranging public health act.

In France in the nineteenth century, unlike in Belgium, England and Italy, there were no doctors specializing in the field of hygiene and public health. It is true that certain specialisms and titles overlapped somewhat with that area: there were doctors who specialized in epidemics, the care of the poor and the treatment of prostitutes, for example. However, their main activities were concentrated elsewhere, in hospitals or private practice and sometimes in higher education. Thus it is hardly surprising that the subjects chosen by medical students for their doctoral theses in the faculty of medicine in Paris should reveal little interest in the area.[3] On the other hand, it is surprising that a consultative institution such as the National Academy of Medicine paid little attention to hygiene and, in particular, to the conquest of water.

However, while the Academy of Medicine was concerned mainly with clinical medicine and pathology, the international conferences on hygiene extended their role to observation and expert appraisement; they were thus justifying and paving the way for the decisions taken by the political authorities.

Finally, at city, department or regional level, health councils relayed their observations made on the ground to the adjoining capital and implemented the directives and circulars that they received in return.

The French Academy of Medicine: observation and consultation

To judge from its *Bulletin* which recorded each of its weekly meetings, the National Academy of Medicine would appear to have taken very little interest in public health and the conquest of water. The Academy preferred to concentrate its attention on the analysis of waterborne diseases.

Stirred up by medical theories and the fear of cholera, the debates of the academicians concentrated mainly on this disease and, as a secondary consideration, on typhoid fever, particularly after the cholera vibrio and Ebert's bacillus had been isolated in 1880. However, apart from investigating a few 'resurgences' of typhoid and cholera, it was less the nature of germ-free water or the drainage of dirty water than problems relating to sera and vaccines that dominated debate in this august body.

Nevertheless, even though small in number, studies relating to sanitation were by no means exceptional. Thus in 1891, Dr E. Mauriac, a member of the Central Hygiene Council of the Gironde, presented several studies on the pollution of the subsoil in the Bordeaux area by cesspits that were immediately declared by the Academy to be 'publications that will oblige the public authorities to adopt those recommendations with respect to hygiene that are recognized as useful.'[4]

Indeed, the main function of the Academy's work lay in producing reports of this kind, which offered the authorities advice on matters of medicine and, more rarely, of public health. Certain in its own knowledge, particularly after the Pasteurian revolution, the Academy spoke derisively of 'the proverbial insalubrity'[5] of the port of Toulon. Thus, according to the rapporteur,[6] it adopted a resolution in favour of 'giving the population as pure a water supply as possible' in order to prevent the return of cholera.

If the need arose, it did not restrict itself to formal advice but expressed its opinions firmly. For example, it condemned 'the very unsatisfactory functioning of the water supply [which] puts at risk the good health of the population of Paris'.[7]

The International Conferences on Hygiene: expert assessment and decision making

'The conquest of water' gave rise to an avalanche of papers, debate and discussion among engineers, doctors, lawyers and administrators, who were all, more or less, hygienists. During the second half of the nineteenth century, major international conferences on hygiene were held in the leading European capitals: Paris, The Hague, Brussels, Vienna, Berlin and Budapest. The British, Germans and French were the main contributors to these conferences. The range of issues addressed was very extensive and virtually no area was ignored. The scientific definition of pure water gave rise to scholarly analyses which had repercussions for the level of sanitation and its overall cost. As soon as the criteria had been adopted, the need was felt for a set of administrative regulations common to all the member states. Scientific norms were paralleled by the social extension of the phenomenon. If pure water was such an undeniable benefit, why should working-class and rural homes be deprived of it? How was it possible to avoid making municipal water rates compulsory? The need for legislation on public health and sanitation was becoming increasingly evident. On this point, the Italian Corradi (from Pavia) was in agreement with his colleague, the Belgian doctor J. Crocq, at the conference held in The Hague in 1884.

The advantages and disadvantages of the combined and separate systems were debated for a long time, as much by Professor Busing of the Berlin School of Advanced Technology as by the Italian engineer D. Sparato, professor at the School of Applied Engineering in Rome, or Doctor Pacchiotti of Turin (1884 and 1903). The spreading of sewage in order that it should be recycled also gave rise to much controversy. Some French hygienists were favourably inclined, while others such as the English doctor Drysdale recommended caution and circumspection. Each used his own example and that of the town or city where he worked. Some also embarked upon comparison, either on a national or even international scale, in particular certain Frenchmen such as Dr Ed. Imbeaux and the chief engineer F. Launay. In about 1900, the question of the purification and filtration of water intended for drinking and cooking was brought up several times for discussion. Hygienists working in the large industrial regions were particularly sensitive on this awkward question: these included the German J. König, professor at the Münster agrochemical testing station as well as the French doctors Henrot (Reims) and Regnard (Paris).

The harm caused by the typhoid and cholera bacteria was clearly strongly condemned, and the engineers, like the doctors, insisted vigorously on the need to monitor water by means of repeated analyses (two, three of four per year, and even more for populations to be given privileged treatment, such as soldiers). The Englishman F. Clernow denounced 'The spread of cholera by water' (Budapest Conference, 1894); the German F. Hueppe wondered about the origin and spread of infectious diseases transmitted by water in substandard supply networks (Vienna Conference, 1887); the Frenchmen J. Bertillon, Dubousquet-La Borderie and Durand-Claye 'dissected' the statistics on epidemics (the cases of Paris and Saint-Ouen) in order to demonstrate the social value of installing a supply of pure water.

Finally, the hygienists gathered at the conferences of The Hague (1884), Vienna (1887) and Berlin (1888) demonstrated their concern with education in order that the principles and standards of hygiene should be diffused as widely as possible. Whatever their speciality or nationality, they were all very aware that, if the nation states wanted to protect their children, it was their duty to make the population healthy and strong and to avoid the procession of avoidable diseases. They all expounded their own creed, declaimed their speechs and celebrated the advent of a clean, clear and luminous 'kingdom' from which all germs would be banished. Their wishes were often heard, despite the problems of all kinds to which they gave rise. In France, the teaching of hygiene was introduced into state primary schools two years after their foundation in 1883. Again in France, Deligny (the creator of the famous Paris swimming pool) proposed in 1889 that the sanitary standards laid down twenty years earlier by the Committee on Unhealthy Dwellings in Paris should be extended to cover the whole of the country: 'Legal measures should be taken to authorize each parish, on the advice of the hygiene commissioners, to make subscription to the municipal water supply compulsory for each inhabited dwelling, to determine the minimum rate in accordance with the available water resources and to fix the rate of subscription.'[8]

The wishes of 1889 became law in France after the great Hygiene Act of 1902. On the recommendation of a famous member of the conference who, in 1889, had declared himself in favour of legal compulsion,[9] it was even, from 1903 onwards, transformed into a budgetary reality.

Because their horizons extended beyond the bodies of individuals to encompass the whole of society, these hygienists, engineers, architects and construction and sanitation engineers were the founding fathers of our industrial society, a society which devours people, energy and water, but which is also avid for standards by which to guide its actions and to measure its performance.

Operating at the junction between knowledge and power, these experts in hygiene forged close links between the notions of order, cleanliness and hygiene and in their practices made little distinction between them. And yet, they 'were notions belonging to different orders and which are not necessarily linked, either historically or in most societies'.[10]

Thus it is essential to recognize the specificity of modern Western medicine, in which the notion of cleanliness has been linked and even confused with those of hygiene, prophylaxis and asepsis. The only explanation for this confusion lies in the importance accorded to germs and the mechanisms of contamination.

In fact, it suited the purposes of science and the ruling classes. As soon as uncleanliness was presented as 'an offence against orderliness',[11] it became a shameful superstition, damaging to health, in short an indication of peasant barbarousness if not of mental backwardness. Thus Lepelletier of the Sarthe region felt extreme disgust when, in 1853, he stopped at the edge of the fountain in Sainte-Anne d'Auray:

The dirt surrounding the fountain and the filth that gives its water a greenish, miry appearance do not prevent pilgrims from plunging their arms into this more or less wholesome fountain and then lifting them up in order to drench their entire bodies in this disgusting liquid.[12]

In about 1900, Guillotin de Corson expressed himself in a similar vein:

Not far from Rochefort in the parish of Limerzel are the chapel and fountain of Saint-Clair; blind people make pilgrimages to this place in order to recover their sight; however, those who wash themselves in the water of this fountain, without piety, immediately do damage to their eyes.[13]

And in 1942, the French doctor Pierre Lazerges condemned this practice in the name of science: 'This popular belief no doubt explains the cases of conjunctivitis contracted after washing in water contaminated by other pilgrims.'[14]

Throughout the period between 1770 and 1940, the notion of the purity of water evolved. Until about 1850, so-called scientific knowledge tallied more or less with popular belief: water was suspect and dangerous when it was stagnant or full of decaying matter, in other words when it was noxious. On the other hand, clear running water – a symbol of life and continuity – was considered beneficial and healthy because it helped to drive away miasmas and waste.

Thus when an authentically scientific, and more particularly microbiological perception replaced, around 1880, a vision that gave meaning to forms, the old symbolism was relegated by scientists to the status of ancient superstition which was not only obsolete but, more importantly, damaging to health.

Thus there emerged a conflict between a tradition, now condemned, that was heavily tinged with religiosity and in which water, like health, was a gift from heaven and the invisible science of hygiene which was alone capable of deciding whether apparently 'innocent' water was indeed as harmless as its appearance suggested.

The 'health record' of water, or the French case

In their anxiety to preserve or to improve the population's health, an essential precondition for the strength of the nation, politicians of the nineteenth century, heirs to a tradition of centralization, did everything within their power to superimpose a hierarchy of health districts onto the hierarchy of administrative districts.

Hastened by the devastating outbreaks of cholera, the health commissions and specialist councils provided for by the law of 1823 were gradually established, even before the decree of 1848 made them more general, extending them to the cantonal, or district level.[15] The members of these commissions, recruited among scientists, health practitioners, administrators and technical experts, were supposed to be loyal cogs in the political machine, transmitting information and implementing directives.

Assisted by the legal system and the gendarmerie and supported by schoolteachers and (in part) by the priesthood, hygienists and administrators fought side by side to enforce health regulations. They were helped in their efforts by the prefects, who were often supported by mayors who were given wider powers in these matters by the Municipal Act of 1884.

The penal code had been made harsher; any person convicted of destroying, damaging or defacing an aqueduct, water main or fountain faced a prison sentence of between one month and two years and a fine of between 100 and 500 francs. And in his municipal decree of 1 May 1849, the mayor of Grand-Quevilly stated that it was his intention to 'prevent the laundry owners in this parish from allowing dirty water from their laundries to flow in the streets' since 'the measure that I have taken is all the more necessary since the cholera outbreak of 1832 claimed many victims in one of the streets through which flows the water from one of these laundries'.[16]

The desire for revenge after the Franco-Prussian War of 1870, the losses sustained in the First World War and the falling birth rate gave rise to a concern with good public health and made more urgent the need to protect by statute the supply and drainage networks. Thus in his circular of 2 May 1922, the French minister for hygiene, social assistance and national insurance wrote to the prefects:

The patriotic struggle against the depopulation of France has many aspects; without losing sight of the overall aim, it can be broken down into a series of concordant and converging measures. . . . Foremost among the responsibilities that fall to the municipalities must be the safety of drinking water; it is impossible to stress too strongly the importance of ensuring the purity of such water, either by collecting it from carefully chosen sources, by improving those which may already exist, which can be done by organizing and providing effective protection for a scientifically determined catchment area, or by purifying water which, despite all efforts, may become a hazard to health.[17]

The public funds allocated to French municipalities from 1903 onwards were intended to ensure that they had an adequate supply of drinking water and to make up for their backwardness in this respect relative to Germany and Britain. The municipalities that benefited were those which had supplied the Consultative Committee on Public Hygiene with a thick dossier based on hydrogeological, chemical and bacteriological surveys.[18] Between 1902[19] and 1925, the statutory regulations governing drinking water were made stricter and more precise.

The number of analyses conducted annually was determined on the basis of the population of the area in question:[20] parishes with fewer than 3,000 inhabitants had to carry out one analysis a year, those with between 3,000 and 5,000 two, those with between 5,000 and 10,000 four, and so on. This was for reasons of public health, since the more densely populated an area is, the greater are the risks of pollution, but also for financial reasons: the number of analyses that could be paid for out of municipal budgets was strictly proportional to the amount of tax raised in the area in question.

Nevertheless, whatever the number of inhabitants involved, certain socio-sanitary standards, derived from studies carried out by scientists, technicians and engineers, were laid down in the legislation and in the statutory regulations that were put in place between 1900 and 1930. Thus the minimum amount of water to be supplied was assessed at 100 litres per day and per inhabitant; this figure was fixed by the scientific world in about 1860 and was well below the 1,000 litres recommended by the International Conferences on Hygiene between 1880 and 1900. Similarly, the hydrogeological, chemical and bacteriological surveys had to be conducted in accordance with very precise instructions and techniques and with all the attention to detail required by the installation of a new type of water supply.

When the hydrogeological survey produced favourable results, the established procedure was to carry out the chemical survey, followed by the bacteriological survey. Once again, it was the purity of the water that was at stake, and every effort was made to penetrate its secrets by means of proven scientific methods. A series of chemical and bacteriological examinations was conducted on samples taken at two different periods of the year (after a dry period and after a rainy period) in order to determine the possible effect of the flow on the quality of the water. These samples were of course taken with the utmost care. The analyses were entrusted to laboratories approved by the Ministry and included at the very least the following determinations:

1 Physical examinations: limpidity, colour, temperature, odour (hydrogen sulphide) and whenever possible electrical resistivity (ascertained at the point of emergence).
2 Chemical examinations: quantitative analysis of organic matter, detection of ammonia, ammonia salts and nitrites, quantitative analysis of nitrates, chlorides, sulphates, alkalinity and iron, hardness.
3 Bacteriological examinations: enumeration and specification of colonies, detection in 100 cubic centimetres of coliform and other bacteria found in faeces and decaying matter; detection of pathogenic species in at least 100 cubic centimetres.

Finally, if naturally pure water could not be obtained, the source chosen had to be the one most suited to effective purification.

The water almanacs

Not content with unravelling the mysteries of pure water, doctors in the West, with the aid of nature experts, planned on several occasions during the nineteenth century to conduct an annual evaluation of waterworks in order to build up a cumulative picture that would benefit the public good.

The idea originated in France. However, as was so often the case, the second edition of the first French yearbook (the 1852 edition which was to have replaced the 1851 edition) took half a century to appear after two abortive attempts by Bechmann in 1892 and Debauve in 1897. . . Other countries, such as the USA (Manual of American Waterworks), the UK (The Water Company Directory), Holland (Overzicht der Waterleidingen) and Germany (Städtische Wasserversorgung), overtook France in the implementation of this major project.

The colossal *Annuaire des eaux de la France pour 1851*, which was the fruit of cooperation between the National Academy of Medicine and the National Society for Agriculture, was intended to serve as 'a catalogue, a sort of shop on which all the industries and professions in any way connected with hydrology will draw abundantly'.[21]

In fact, at that time, only one drainage basin, that of the Seine, had been properly analysed. As a result, the water in the Paris basin came to represent that in the rest of France and became a focus of attention for agronomists and doctors. As an inference from the theory of the atmospheric water cycle, pollution was examined with great care. Researchers compared the amounts of 'putrid matter' and 'solid substances' found in samples of water; the quantities were greater in those taken downstream of Paris than in those taken upstream of the capital. The city's image, already tarnished by Rousseau, was further blackened by this finding. The Bièvre, whose purity had been much appreciated by the Gobelin brothers, was declared to be polluted by the 'domestic needs' of those living along it, by 'a large number of laundries' and in particular by 'the business premises of turners, tawers, tanners, leather workers and dyers and starch factories and bleaching houses.'[22]

Nevertheless, and let there be no mistake about this, the conclusions of nineteenth-century scientists do not in any way presage the alarmist calls of present-day ecologists. They noted with satisfaction the 'slightly laxative' property of the water of the Seine in Paris, thus reflecting the words of a Dutch observer during the Regency period: 'it often gives strangers to the city stomach ache and, in the words of the French, one pays tribute to it.'[23]

At Rouen, scientists observed in 1851 that the waters of the Robec were 'sometimes so dark in colour that it was difficult to believe that cotton garments could be washed in it'. Nevertheless, as we now know, they were right to conclude that 'a night of rest was sufficient [for the Robec] to regain the limpidity and transparency that it has at its source.'[24]

By foreshadowing in this way the self-cleansing capacity of rivers and advocating the construction of a sanitation system for Paris, they declared themselves capable of finding solutions to the pollution problems that they discerned. In 1851, the reviewing and evaluation of water in France was still limited to the Paris basin. It was not until 1903 that a nationwide assessment was made.[25] This provided an inventory of resources that was extremely beneficial to economic activity and health in France. It led to a rise in the standard of living, a concomitant rise in hygiene standards and a decline in certain diseases. The few water directories published in the nineteenth century are thus the direct forebears of our present-day institutions: in the case of France, the Water Act of 1964, the inter-ministerial water commission and the organization of the French Water Board into agencies based on river catchment areas; in the case of Britain, the setting up of the Water Authorities in 1973 and the autonomous regional drainage committees.

The purity of water was now more than ever at the heart of the debates. Among the scientific and political elite, it was the subject of rigorous care and attention. Henceforth, particularly towards the end of the century, *scientific* knowledge of water became part of a vast project of social and cultural change inspired by the utopian visions of the time.

This 'water mania' was reflected in the 'sanitary engineering' advocated by doctors, technicians and engineers. They peopled the countryside and the home with new devices intended to drive out impurity and wash out dirt. Marianne, the symbol of this conquest of water, now adorned the fountains of the French Republic, replacing the Melusinas of mythology and the saints of medieval times.

The use of domestic water in boilers from 1900 onwards provided competition for the public wash-houses and consigned large traditional washes to the museum. The use of new sanitary equipment, such as baths, bidets and WCs, modified the interior of some dwellings and led to the development of new rooms reserved solely for the practices associated with personal hygiene.

Finally, hygienism, which was concerned with an environment, and Pasteurism, which dealt with infectious diseases, both helped to increase the importance attached to water; the subject of scientific and technological advances, it benefited from an aura of innovation, but also retained for a long time its ancient virtues.

And because its scientific and technical value was now better known, fresh water became the prey of a new power: that of money. It had for a long time been regarded as a gift from Heaven, but it now became a product controlled by human activity. Afflicted by industrial 'fever', it eventually became a 'gilt-edged investment'. It was used in vast quantities by industry, from textiles to pharmaceuticals and iron and steel; consumption by private individuals and city authorities for reasons of hygiene and cleanliness also increased. Finally, fresh water was carving out a kingdom for itself, a kingdom both visible and invisible, subterranean and manifest, public and private, intimate and social. A water-devouring economy was gradually set in place. However, whether in 1770, 1870 or 1930, there was very rarely the feeling, unlike today, that this natual resource was being wasted and depleted.

In reality, it was a period that displayed rational euphoria. Its 'faith' was based on a materialistic revolution that was believed to be without precedent, a revolution that heralded unlimited growth and prosperity. It rested in particular on breakthroughs in scientific, technical and medical knowledge. By bringing into play bodily and natural mechanisms, this new knowledge made it possible to find effective solutions for the problems raised by the increasing consumption and pollution of fresh water.

With the aid of various techniques, ranging from the very simple (household filters) to the most advanced (chlorination, use of ozone), the ideologues of progress were always able to suggest a remedy which would eliminate impurities and ensure a carefree future. However, in order to triumph over society and to achieve its full effect, the conquest of water needed 'drive belts' through which the new knowledge could be transmitted. These it found in the press, the schools and the hospitals, and it was through these institutions that the project became reality.

PART II

Mass Diffusion

In the course of the nineteenth century, the hygienist image displaced representations derived from an ancient pagan culture which had been slowly vanquished by Christianity. The cult of water was secularized and came to permeate everyday life. This process of diffusion was carried out through various channels. Their common characteristic was the emphasis they placed on the scientific aspects of water and its socio-sanitary code. Hardly had the new knowledge left the hands of scientists than it was elevated to the status of model: it was proclaimed from the tops of fountains by statues of Marianne; the combined town halls and wash-houses bore witness to it; handbooks of etiquette dwelt on it; and advertising in newspapers and weeklies meted it out repeatedly. Finally the secular primary schools, the army and the hospitals taught it and tried to put it into practice. We shall use the example of France to examine this process of diffusion.

4

The power of the press

'Signs and street cries [. . .] are the original forms of all publicity.'[1] Street cries were the forerunners of advertisements, quack doctors presaged pharmaceutical advertising and the doctor Théophraste Renaudot was the inventor of economic information. Indeed, 'everyone can see that our publicity still consists of street cries and quack's patter.'[2] Nevertheless, the emergence of strictly commercial advertising is a much more recent phenomenon. It can be located between 1788, the year in which *The Times*, the first vehicle for large-scale advertising, first appeared, and 1845, when Charles Duveyrier, a follower of Saint-Simon, established the Société Générale des Annonces, the first large-scale publicity company. To be more exact, in the case of France, it was Emile de Girardin who, in 1836, expounded his ideas on advertisements when, in launching his newspaper *La Presse*, he was induced to explain the merits of commercial publicity.

His method, as expounded in the brochure that accompanied the launch of *La Presse*, was to channel the flow of advertisements in order to bring about a reduction in the price of subscriptions, which would in turn enlarge the readership of his newspaper. He redefined the advertisement. It was no longer a parasite, but, allocated to its own section, it guaranteed the financial health of the newspaper and its informative and promotional role. It was from this exchange between information and advertising that the modern press was born, since it was advertising that created a vast market for daily newspapers. From that time onwards, needs were induced, tastes were chanelled and prestige came within the scope of the 'man in the street', particularly in the key sector of hygiene and beauty, which was the first to be exploited and from which the highest profits were to be made, as they still are today. Balzac was not mistaken when, through his character César Birotteau, he attempted to describe the rise of the capitalist system. Birotteau's 'cephalic oil, based on principles laid down by the Academy of Sciences, produces that important result, much sought after by the ancient Greeks and Romans and the Nordic peoples to whom hair was precious.'[3] Fired by his dream of destroying his colleague Macassar, Birotteau uses the sensationalism of his Prospectus and the glitter of the 'princely' ball that marks his apogee to do battle with his time and manufactures his product before identifying his customers. Based on calculation, flair and an act of faith, his notion of advertising develops into a 'volontarist and magical vision. . . which carries within itself the splendours and dangers of the Promethean myth.'[4] He attempts to bring luxury within the reach of those for whom it represents advancement. He uses the scientific argument of innovation, but cleverly links it to a long tradition, and appeals finally to the fantasy of the symbolic and sexual power that, since Samson, has been associated with the beauty and length of hair. 'The decor of the fancy goods window vindicates audacity by making it part of a tradition, of an established culture.'[5] In fact, all advertisements can be interpreted according to a dual code: that of the obvious usefulness of the product in question and that of the impulses or fears that it symbolizes. Behind the reasonable argument and functional illustration, 'advertising addresses the secret, though universal society of the unconscious.'[6]

In the nineteenth century, the take-off of the industrialized economy and a period of stable currency led to a universal, though socially differentiated, rise in the standard of living. As a result, gaps appeared in the market and new products, the need for which was often created, took their place in the shop windows of the West. This was true of the health and sanitation industry; as the share of household budgets allocated to food tended to diminish as families climbed the social ladder, a market was created which advanced by leaps and bounds. It turned pleasure, convenience, comfort and aesthetics into commodities; physical appearance and body odours were directly implicated, as were underclothes, though more peripherally.

Thanks largely to Emile de Girardin, advertising accounted for an increasingly large share of newspapers' revenue. The press became a sort of relay system for the mass diffusion of a new relationship between water and the body, in which the old notion of cleanliness, based on decency and concern with appearances, was combined with a scientifically based notion of hygiene and the virtually eternal fear of intimations of mortality and the desire to be protected from them.

L'Illustration: The body and the bourgeoisie

Established in 1843 by three disciples of Saint-Simon, *L'Illustration* appeared until 1944.[7] It was a weekly read initially by the moderate and subsequently by the conservative middle classes in the towns and cities.

This advertisement, which appeared in *L'Illustration*, shows the 'sanitary trilogy' of bath, washbasin and bidet installed in close proximity to each other in a specific place, the bathroom.
(Photograph: © D.R.)

It was distinguished by its high technical quality, still unequalled today. By about 1930 it had become old-fashioned and much more politicized; its popularity grew considerably during this period until its support for the Vichy regime led to its disappearance. Around the turn of the century, it had a circulation of about 50,000 copies, which rose to 200,000 in 1910 and to 400,000 during the First World War, after having been taken in hand by René Baschet. It was expensive (5 Francs a copy in 1931; 185 Francs for the annual subscription in 1935), exclusive (it would be considered precious, almost ostentatious today), printed on luxury paper and carried magnificent, virtually unequalled illustrations.

The main emphasis in the advertisements that it carried for hygiene products was on luxury items (toilet waters, famous perfumes, bathrooms) and sophisticated tastes (elegant spas, renowned mineral waters; beard care occupied a leading place in the advertising columns).

L'Illustration was characterized by a profusion of advertisements and products. The trappings of luxury and pleasure involved all aspects of bodily cleanliness, both internal and external. The bourgeois body was thus immersed, pampered and titillated; it was a social asset to be preserved and made fruitful. Advertisers' techniques were deployed to capture the reader's attention; these included special characters with boxes, illustrations (photographs after 1869) and some practical information, such as the address of the retailer or manufacturer and, more rarely, the price. The advertised product was thus intended to capture the attention and arouse the desire of a well-to-do population, susceptible to the attractions of water. It was possible here to glimpse the desired balance between a bourgeois body giving itself up to inspection in all its chaste glory and desire for technical and aesthetic efficiency, and the water that imbued that body with its secret charms and beneficial powers.

The arguments that are developed in these advertisements combine to ensure the safety of the reader/voyeur. This is achieved in several ways. Some of the arguments appeal to logic, others to knowledge or physical and moral guarantees. However, the most important one is based on the direct and tangible technical efficiency of the advertised products. Two subsets can be discerned here, one based on medical and the other on more discreet scientific arguments.

The reference to scientific knowledge reflects the prevailing belief in its 'virtues' and beneficial powers. More particularly, the reference to medicine expresses the trust now invested in doctors, who had become indispensable members of society to whom control of the body, and thus of its relationship with water, had been entrusted. The dominant image in the sales arguments is that of the father and, in second place, that of the scientist, under the name of Pasteur.

The advertising was based not so much on value for money, which was a virtually unknown concept at that time, but rather on sensitivity and make-believe: the aestheticism of the bazaar, prudish sensuality, a taste for luxury, or what passed for luxury. The image of women was a pink, perfumed one, while decent men were concerned only with their face (teeth, hair and beard).

Between 1830 and 1940, the importance of the various arguments changed: the theme of efficiency remained pre-eminent throughout this period, even though it stagnated after 1914. The sociability theme was also sustained, but its components changed: pleasure was emphasized during the '*années folles*', as was the taste for luxury, while the following period, until 1920, marked a very understandable trough caused by the economic crisis that hardly encouraged the glorification of the body.

From the Second Empire until 1940, it was the thermal spa which carried the torch of bodily hygiene and the relationship of the bourgeoisie to water. This had not always been the case: during the Second Empire, this leading role was played by soap and toilet water; after 1900, it was played by face and hair care, while after the First World War dental care gained the ascendancy.

Advertising of this period – still so close and yet utterly remote from our time – obviously used or took over 'events' in order to promote products. In 1880, 'Dr Delabarre's Toothpastes' were touting their 'Oriental Water', their 'Oriental Powder' and their 'Oriental Paste'. In 1915, 'pure water', that is mineral water, had to be reserved for soldiers. In 1943, a manufacturer of beauty products announced that he had given 200,000 francs to assist prisoners of war. Appeals to the national fibre were thus not ruled out, any more than awareness of the nation's history and of the empire which can be discerned in the names and images of certain products: a moustachioed Gaul with long tresses was used to promote mineral water, 'Suez water' was famous between 1870 and 1900 (see photograph), Congo soap combined the aquatic allegory of the great river with the charms of a beautiful negress, and, as has been noted, during the Third Republic Marianne found herself hoisted to the top of many fountains in France.

Possession and appearance were the two imperatives placed before the middle-class readers of *L'Illustration:* the possession and enjoyment of health, thanks to scientific knowledge, based on medicine, and appearance in society displaying the signs of distinction. On the one hand, there was the universe of the doctor – the traditional investigator – of medicinal and then of mineral waters; on the other, the world of the bourgeois who managed his own body as a good *paterfamilias* and immovable pillar of the social order.

A bourgeois ideology emerged, upright and serene, imbued with its own importance. Its efforts to legitimize its place and role in society reflected an overwhelming desire to dominate and be master of a body that was a source of questioning, pleasure and anxiety. Both text and illustrations revealed a domesticated and medicalized nature, like an animal that had been tamed and taught to sit and beg; hotels were full to overflowing with stucco and blind arches, parks were laid out geometrically and bourgeois patients taking the waters nonchalantly displayed their ill health. Finally, the images of the spring led to the appearance of a technological 'fountain of youth', firmly under the control of man, who was able in this way to assert his power over nature.

Even better was an advertisement from 1910 for a lemonade, the forerunner of Vichy-Saint-Yorre, which was described as 'exquisite, natural and pure'. The advertisement, which depicts a young peasant girl returning from market with two bottles in her basket, combines the various themes used to symbolize purity: a virginal spring flowing out of the ground, the countryside and the sacred liquid itself, imprisoned in its bottle under the designation 'Vichy Lemonade', a symbol of man's domination of nature. However, the pseudo-rural image did not hold sway unchallenged: a 1936 advertisement for Perrier water depicted an engaging lady tennis player, a worthy rival to Suzanne Lenglen, drinking, against the backdrop of a tennis court, an 'elegant, sporting water', a future 'champagne of table waters'. Finally, in 1943, the publicity for Baril toothpaste juxtaposed a young woman and some shells with the following slogan: 'a smile of pearl and coral'.

Thus it was that an entire mythology was constructed: Hygeia and Panacea, priestesses and daughters of Chronos, took their place in the bourgeois Olympia in the guise of the peasant girl and sportswoman. These images, brought back to life from the *Dea mater* of antiquity and the *vix naturae* so dear to Galen, served as a foil for the interests of capital; hitching up their skirts, they went in pursuit of the middle-class consumer.

The illustrated supplement of the *Petit Journal*: the working classes and their bodies

Established in 1863, twenty years after *L'Illustration*, *Le Petit Journal* was a popular daily newspaper. Its illustrated supplement first appeared in 1890.[8] In 1864, it had a circulation of 245,000 copies, rising twenty years later to almost a million and declining to 800,000 in 1914 as a result of competition from *Le Petit Parisien*. In the 1900s, it had about 18,000 agents, which enabled it to consolidate its position in the rural areas and the provinces where a good half of its readers were located.

In 1911, its founder, Jules Léon, described its readership thus:

It is made up for the most part of the provincial lower middle classes, people living on small private incomes, manual workers, tradesmen and caretakers; they are all people with little education, limited ambitions, in a more or less dependent situation. . . . They are honest, peaceable, settled, thrifty, hard working and utterly averse to adventure of any kind.[9]

Zola, who worked on *Le Petit Journal* for a few months in 1865, even spoke of the shepherds who read it while watching their flocks!

The illustrated supplement was aimed at a working-class audience. Its cover price is evidence of this (5 centimes in 1900, 1 sou 50 centimes in 1934), as is the poor quality of the paper, colours and illustrations.[10] Its audience is also reflected in the products advertised in it. Luxury items, such as toilet water, bathrooms and thermal spas, are almost completely missing. On the other hand, cheap basic products such as soap, shaving cream, razors and shampoo, which had become essential to the cult of cleanliness, are advertised as frequently as in *L'Illustration*. The mineral waters intended for middle-class consumers are replaced by a much cheaper chemical product for mineralizing tap water.

The social status of the readership is also reflected in the much narrower range of products on offer and the size of the advertising spaces, which are nine times smaller than in *L'Illustration*. Boxing of the text replaces the pictures, and drawings the photography that appeared in *L'Illustration* from 1869 onwards.

However, despite these undeniable differences, there is a similarity in the sales gambits used in these advertisements which suggests that they might have been responsible for spreading into the working classes, or at least the 'better' elements thereof, a significance based on the middle-class model. Indeed, in both these weeklies the efficiency argument seems to be fundamental: it is used in 78 per cent of the advertisements in *L'Illustration* and in 89 per cent of those in the illustrated supplement. Medical guarantees predominate: they feature in 60 per cent of the advertisements in *L'Illustration* and in 67 per cent of those in the supplement. As far as scientific guarantees are concerned, they are twice as common in the working-class weekly as in the middle-class publication.

Finally, in both the supplement and in *Le Petit Journal*, the advertisers appeal essentially to the reader's sense of imagination: to his faith in science, his belief in progress and his desire to escape from the wear and tear of everyday life. They never allow him to operate rationally. The advertisements usually carry no mention of the price and composition of the product, which makes it impossible to assess whether they offer value for money.

The same cultural model would thus seem to have emerged, differentiated by social class. Aesthetics, luxury and pleasure were the prerogative of the middle-class weekly. On the other hand, if the evolution of the two main sales themes – security and sociability – between the second half of the nineteenth century and the eve of the Second World War is examined, the same pattern of development emerges for both publications.

A shared concern with efficiency

Whether one is making an overall assessment or analysing a set of circumstances, social diversity and cultural unity would appear to be compatible notions. The search for efficiency differed only in the forms that it took: if soap was the 'standard bearer' for bodily hygiene among the working classes, the spa fulfilled the same function for the bourgeoisie. Each, in their own way, was fulfilling a need that was to a large extent induced, the need for a clean, healthy body, free from any physical contamination, whether visible or invisible. Internal cleansing with spa water was reserved for the bourgeoisie, while the working classes had to make do with using soap for external cleansing.

In fact, spas did occasionally feature in the popular weekly between 1884 and 1885, but they then disappeared entirely, while there was no advertising at all for toilet waters (see table). Except for a few brands of toothpaste, there were thus very few products advertised in both the bourgeois and working-class publication, although it is possible to detect certain changes in the course of the period in question.

| | Advertisements | | |
Publication	Aesthetic argument	Reference to luxury	Pleasure and sensuality
L'Illustration	29%	15%	28.6%
Illustrated supplement of *Le Petit Journal*	22%	0%	14.8%

After a period of twenty years, the sanitary products involving the use of water that were advertised in the popular newspaper followed the model put forward by the middle-class weekly. This was true for soap, hair and face care and mineral water. On the other hand, sanitary equipment and modern conveniences such as shower rooms, bathrooms, showers, baths and water heaters remained the prerogative of the middle classes.

The unifying element in these two representations of water, one bourgeois, the other working class, lay finally in the explicit desire for a cleansing of the body intended to purify, maintain and revive it. Cleanliness, even when expressed in the vocabulary of hygiene, always had a positive, even laudatory connotation. It was a struggle for life, a strategy based on efficient, operational methods of administering an inheritance. The body's relationship with water was perceived as a return to its source in an attempt to ward off ill fortune and to kill the germs, doubts and pollution received and given off by the body.

Only fresh, pure running water was in evidence, since it was only that vibrant symbolism that could underwrite the sale of products intended to eliminate waste, fatigue, ageing and death. It was indeed life itself that was on sale. Stagnant or turbulent water thus had no part to play.

In both weeklies, the advertisers had obviously allocated specific roles to certain parts of a modest body, as required by traditional handbooks on politeness. The bust and hands of a woman symbolized femininity: humidity, warmth and fecundity, grace and cleanliness, with just a hint of seduction and anticipated sin! The hair of a man's face, associated with the cutting blade of a razor, depicted the 'strong sex': power, muscular strength, use of the tool and the vigorous growth of hair without which he would cease to be a man.

Children's bodies made only rare appearances, in advertisements for toothpaste (1910) or toilet water (1936). This is a sure sign that advertisers were directing their energies for the most part at the adult world and, in particular, at men, who controlled the purse strings. It probably also reflects a deep-seated puritanical belief that children should not be used to advertise products.

Despite the technical nature of the products being advertised, the bodies were always young, beautiful, clean and fresh, as if they had just emerged, purified, from the sea. The only image of imperfection ever used was that of male baldness. However, this was simply to show all the more clearly the effectiveness of the advertised treatment and to prove that the remedy was capable of turning back the clock.

Similarly, the megalomaniac multiplicity of 'virtues' and qualities with which mineral waters were imbued owed as much to a nostalgia for paradise lost or the Panacea of antiquity as to the medicalization of the body and of water, which was taken to extremes during the triumph of Pasteurism. It was for this reason that glimpses of anxiety, obsession even, with an always imminent danger that had to be eliminated, tended to emerge from behind the radiant descriptions. In 1884, a mouthwash was described as 'anti-epidemic'; in 1885, a household filter was said to provide 'microbe-free water'. By about 1920–5, the principal adjective used to describe soap and mouthwashes was 'antiseptic'.

Thus the explicit reference to Pasteurian medicine in the struggle against infectious diseases readily became part of a much more general design: that of Promethean man battling against suffering, disease, death and the forces of darkness.

Thus the same song, whether heard as a dull murmur or intoned quite openly, arose from the advertisements; it was a monumental incantation intended to stave off the passage of time, banish the shadow of death and gain wellbeing, happiness and health. Its only originality lay in its scientific and technical formulation and in the medium by which it was diffused: newsprint, illustrations and photography had replaced stone and stained glass. Nevertheless, the ancient symbolism of water was still present in order to promote a 'healing' product: the images of the body as capital, the body as machine and the body in its wild state were articulated through their relationship with water, the gift of God or of nature. Through the mediation of money, water – a product created by man – gave him extra life and, in separating out the winners, provided the right response to the 'decisive question'!

Advertising in nineteenth-century France was thus far from being innovative. It exploited a substance fundamental to most religions known to man, whether they originated in Australia, North America or the Middle East. Secondly, as Emile Durkheim understood, it made the substance into a real cult again.[11] Through a set of ritual practices, it invited the public to draw nearer to the sacred world and to develop positive, bilateral relationships with it. Finally, the life-giving aspect of the symbolism of water was widely used by advertisers in both the working-class and bourgeois publications. Its 'rehabilitation' even constituted the basis of their sales pitch.

The discourse that emerges from this repetitive publicity makes extensive use of 'rites of regeneration'[12] intended to eliminate any discontinuity between the cosmos and the body and to incorporate a much older cultural attitude into a new unity, based first on hygienism and then on Pasteurism. Thus, to use the terminology of Auguste Comte, the symbolic state and the religious state became part of the scientific state.[13]

Between the second half of the nineteenth century and the 1940s, a new relationship between the body and water emerged in both urban and rural France. Diffused among the population at large by advertising, it was intended to eliminate 'miasmas', to kill 'germs' and to mould those classes considered to be dirty and dangerous in accordance with the received wisdom of the time. The novelty of this wisdom lay in its determinism, its volontarist intentions and its concern with technical efficiency. Nevertheless, it used, and, when it was not subjugating it, rehabilitated an extremely ancient symbolism which afforded glimpses of the desires and anguishes that the ample cloak of science and reason could not succeed in hiding.

In so doing, it fought and forced into retreat or change the old peasant culture that was typical of the Christian West. There is no doubt that advertising – an archetypically urban phenomenon – contributed 'greatly to the diffusion of the most prestigious models of bourgeois behaviour, the adoption of which, even though only partial and inept, always led to increased consumption'.[14]

5

The role of the hospital

The 'healing machines'

In the second half of the eighteenth century, 'the doctor became the great counsellor and expert if not in the art of managing then at least in that of observing, correcting and improving society and maintaining it in a permanent state of good health.'[1] Even more than his prestige as a therapeutist, his role as hygienist gave him not only the right but also the duty to intervene with authority in institutions and organizations such as cemeteries, churches, prisons, ships and general hospitals that posed health problems. The intervention might be positive, for example, when it involved the implementation of a 'concerted therapeutic strategy',[2] or negative, if it was aimed at eliminating those factors that made hospitals dangerous for staff and patients.

Hospitals were institutions for the treatment of disease rather than for the provision of welfare, and, by virtue of their layout and fittings, the circulation of air and water, the disposal of waste and the washing of linen and the wards, 'had to function as healing machines'.[3]

The role played by the hospital's water supply was fundamental. Nevertheless, it was not entirely novel. At the height of the Middle Ages and during the modern period, all such communal institutions – prisons, ships, monasteries and colleges – had shown great interest in the installation of one or more 'water lines'. In the seventeenth century, the Hôpital Saint-Louis, in order to escape the plague, had a large reservoir built; at the beginning of the eighteenth century they changed the route taken by the supply pipes. The pumping system that supplied water to the various parts of the hospital was installed in 1722 by Davesne, the master locksmith at the Hôtel-Dieu.[4]

Similarly, the water supply in the old seventeenth century pesthouses in Marseilles constituted a serious and much discussed problem; thus in 1653 a canal was dug in order that water could be supplied from an 'old fountain'.[5]

Even though priority in the second half of the eighteenth century — and for more than a hundred years afterwards — was given to the 'circulation of air', the water supply continued to worry those in charge of hospitals, who were anxious to create a 'new spatial order', to make hospitals salubrious, to identify the 'morbid entities' and to turn hospitals from almshouses, or prisons for the destitute, into 'healing machines' for the benefit of society as a whole. Until about 1860, the model established more than a century earlier remained virtually unaltered.

Fact-finding tours, surveys and questionnaires all helped to build up a body of knowledge on which new norms could be based. Thus in the *Report of the Commissioners instructed by the Academy to examine the proposal for a new Hôtel-Dieu* in 1785, it was stated that 'a hospital built in an enlightened century such as ours has to be the product of acquired knowledge; it must bring together all the assistance that advances in the skill of healing can offer for the relief of the sick.'[6] However, this praiseworthy intention was not in any way matched by reality. Even the Hôtel-Dieu in Paris was 'neither comfortable nor salubrious'. Thus:

Saint-Charles ward obtains its daylight only from verandahs; but what daylight and what air! These verandahs are intended for the drying of linen, because the Hôtel-Dieu has no space for clothes lines or drying rooms. These verandahs are thus encumbered with numerous iron clothes driers piled as high as the casement windows in the ward. Greyish wet linen is always hanging there, and it is only through these obstacles that daylight, gloomy and constantly humid, penetrates into the Saint-Charles ward.[7]

Jacques-René Tenon (1724–1816), former chief surgeon at the hospital of La Salpêtrière, was commissioned by the Academy of Sciences to draw up a report on the rebuilding of the Hôtel-Dieu. During the Constituent Assembly of 1789, he was President of the Committee of Salubrity. In his *Demandes relatives aux hôpitaux* (1790–1), he drew up a twenty-eight point questionnaire, five of which related directly to the question of water.[8] He wanted to know whether the hospital 'is situated on a river, or a long way from it; and if it is a long way away, by what means and in what quantities it is supplied with water', and whether the main wards were washed. He requested 'details of the wards, the heated public rest rooms and particularly the lavatories or cesspits' and wondered whether 'only [. . .] close stools' were used. He also requested 'information on the order observed in the distribution of water, in order that there should be an abundant supply everywhere for daily need and in the event of fire'. Finally, he asked for 'details of the showers, steamrooms and bathrooms' and that he should be told 'the number of these rooms, the number of baths and the means used to obtain hot water'.

Even though Tenon was not concerned with the changing or washing of linen and clothes, he was one of the first to conduct a systematic review of the various uses of water in hospitals: for drinking, for washing the floors, for treatment (baths, showers, steamrooms), fire fighting and the disposal of excreta.

The gap between the situations observed and the recommended ideals was not a small one. Thus the Bordeaux hospital, which had 400 beds, 'is situated on a stream known as the Devèse which has little water most of the time and and from which unbearable stenches emanate throughout the entire hospital'. Indeed, 'there are no lavatories for the patients; there is one close-stool for every four beds that are emptied once or twice a day into the Devèse, whether there is any water in it or not. And when the water level rises, the excreta are swept away towards the river.'

The water supply was 'primitive' and did not meet any of the criteria of salubrity laid down by the academicians in Paris: 'There is no water in the building except for a fountain that does not always work; when there is no water, a well is used to supply most of the required water.' Finally, although the main wards were washed, hydrotherapy was unknown in the Bordeaux hospital, as was a hot water system.

This absence of communal hydraulic equipment was fairly common at the end of the eighteenth century. Indeed, welfare for a long time remained the responsibility of the parishes and the balancing of hospital budgets was a virtually never-ending headache. On the eve of the Revolution, hospitals, both large and small, were hardly models of hygiene or cleanliness. Seventy per cent of hospitals − all those which were neither principal or general hospitals at the end of the *ancien régime* − had on average between twenty-five and thirty beds; and 'in forty per cent of cases their capacity is less than ten beds. Very often, "hospitals" were no better than the hovels described by researchers in the eighteenth century which were given over to a helpful old woman looking after a few straw beds and tending the sick at home.'[9]

This was the case at Malestroit, where the hospital was 'an old establishment set up solely for the old and infirm; the building is small, old and decaying.'[10] All the patients were housed in the 'poor ward', which was

unpleasant, uncomfortable and damaging to the health of the poor people because of the bad air and malignant humours that pervade the ward.[11] Some of the bedding on the six old and decrepit beds is a hundred years old and the beds are rarely changed. In 1748, 'Mistress Brandeul, the poor warden at the said hospital, complained that the gowns for both men and women and even the sheets on the bed were completely frayed and worn out by the number of sick people who had been at the hospital'; forty years later, Mistress Fantou 'complained that linen was in very short supply in the hospital.'[12]

From the Age of Enlightenment onwards, some hospital wardens not only denounced this state of decrepitude, but also, at least in the case of the richer and better managed establishments, set about improving the salubrity of their hospitals and installing a water supply. Thus, at the hospital of Provins, Prior Gribauval in 1762 'had a three foot trench dug out in the women's ward, which was damp and thus considered unhealthy for the patients; the trench was filled in with stones in order to purify the ward. He also had a water supply and drainage system installed and then the whole ward was paved with large freestones. The ward was illuminated by windows on the eastern side.'[13]

And when it was not a water supply that was being installed or improved, some hospitals were installing those water filters which came into increasingly widespread use among the well-to-do and enlightened classes during the second half of the eighteenth century. Thus at La Salpêtrière in 1756 'for the use of the pharmacy, a fountain with a sand filter will be installed with a supply of river water that will be adequate for infusions and other drinks for patients, for the formulation of other medicines for internal use and for those chemical processes that require the purest water that it is possible to obtain.'[14]

For a long time, however, the hygiene and cleanliness of most French hospitals and, more specifically, their hydraulic equipment, left a great deal to be desired, even if, about 1830, certain US students expressed their admiration of the Hôtel-Dieu in Paris.[15] Thus the 'fundamental defects' at the Hôtel-Dieu in Lyons condemned by Isidore de Polinière in about 1850,[16] included the insalubrity of the courtyards which 'have openings intended to collect rain and domestic water that are linked to the sewers', the absence of running water (water was carried into the wards by hand) and the lack of baths.

Despite undoubted improvements in hydraulic equipment, 'pure' water remained a precious commodity for a long time, as shown by the wrangles about the rare and expensive 'water lines' that took place in the sixteenth and seventeenth centuries between hospitals, colleges and convents. Thus in 1902 in Croix (Nord), 'the only liquid that comes out of the taps is yellowish, malodorous and brownish that cannot possibly be used even for washing or cleaning the floor; the liquid that emerges from the town's water taps gives off a dreadful stench; there would be grounds for obliging the town not to infect the hospital.'[17] Much worse, before 1884, only five hospitals in Paris had a supply of running water for domestic uses: the Hôtel-Dieu, Bichat, Saint-Louis, Tenon and the Maternity Hospital.[18]

Nevertheless, there had been various phases of major works and modernization in the public hospitals in Paris, mainly between about 1852 and 1890. Reported by Husson and Davenne, analysed by Tollet and studied by Fosseyeux,[19] the purpose of the large-scale works was to introduce hygiene into the construction and arrangement of the buildings and into the wards, thus implementing the formula drawn up by Jules Guadet: 'A hospital has only one aim, which is to heal, and everything must contribute to that aim. The architects must work towards that end as much as the doctors, and no less efficiently.'[20] In the years between 1859 and 1880, hospitals 'aspired to be not merely the place of healing, but a factor in the process.'[21] In support of the same principle, Tollet wrote in 1892: 'I shall finish by recalling the wishes that I have never ceased to express: that the number of hospitals should diminish and their equipment and fittings improve, to the point where they will soon be veritable instruments of healing.'[22] In order to achieve this fundamental aim and to break with the notion of the hospital as a home for the old and infirm,[23] Tollet, as a good technocrat, stated:

In the construction of hospitals, three equally essential conditions must be fulfilled: hygiene, economy in construction and service facilities, the corollary of which is economy of operation. However, these three conditions are frequently in conflict; in implementing them, the hygienist/architect must strive to reconcile this conflict.[24]

Until about 1860, one single architectural procedure was to provide a solution to the problems raised by contagion: isolation. The Tarnier ward (1870) was the most efficient and representative method of protection. However, the discoveries of Pasteur and Koch had demonstrated that the sick person himself was the main centre from which contagious germs emanated. This gave rise to the notion of 'establishing within the hospital not one but several lazarettos linked to the public wards but completely isolated from them'.[25] Thus with the changes that had taken place in medical knowledge, priority was given to the fight against infection: 'See to it that a sick person is never a hazard to his neighbour.'[26] The general salubrity of the hospital was relegated to second place: 'an uncontaminated water supply, the disposal of waste and toilets built in such a way as to be neither dangerous nor inconvenient. . .; then convenience in the economic operation of the general departments'.[27]

It had taken a century (1770–1870) to reach this point. Importance had been attached to the salubrity of hospitals since the projects of Petit (1774), Poyer at the Ile des Cygnes (1786) and Leroy (1777). The separation of the healthy and the sick and the culturally determined dichotomy between the principles of life and death were finally to lead to the notion of the hospital as a place where miasmas and filth would be driven out by the beneficial action of air (aeration), water (for medicinal uses[28] and the disposal of waste) and sun (in the fight against miasmas).

The visits made by Tenon and Coulomb to English hospitals led the Academy of Sciences to propose a system of 'separate yet parallel wards, laid out facing the same direction on several floors'.[29] Clavereau's report

of 1805[30] examined in detail the question of aspect and recommended
that hospitals should be built on the side of a hill where there were
plentiful sources of clean water, since a site of this kind would make
available at the lowest possible cost, that is without the use of hydraulic
machinery, all the water that the hospital would need.

The projects of Duchanoy (1812) and Gour (1832), the schedule for
the construction of the Hôpital Lariboisière (1839), the British
programme (1855) drawn up by the committee chaired by Lord Munk,[31]
the discussions held in 1864 by the Paris Association of Surgery[32]
following the publication of Trélat's treatise[33] and the programme of the
Committee of Health for the French military forces (1872)[34] all
recommended high standards of salubrity and irreproachable cleanliness,
with the exception of the last named, which preferred dry disposal rather
than wet drainage for water closets. Running water was seen as a means
of ensuring proper hygiene standards in hospitals and was intended to
supplement the salubrity achieved by ventilation and isolation. This
applied as much to urinals and water closets as to laundries and linen
rooms. In hospitals, however, water retained its therapeutic qualities.
Although the innovation (particularly after 1860) lay in the hygienic and
preventive uses of water, bathrooms and showers for patients were for a
long time constructed with reference to Vitruvius and eighteenth-
century medical practices.

Thus for reasons connected with the medical theories of the period
1770–1870, the plans for the 'monumental hospitals'[35] placed faith not
only in aeration but also in the hygienic and medical use of water in
order to cure the sick. Nevertheless, according to Tollet, hospitals in
France continued to be built in accordance with an old-fashioned and
mistaken model until about 1875: handsome facades concealed internal
squalor, and the whole design made the cost of each bed very high, in
contrast to the policy adopted in Great Britain.[36] Even in 1910,
Fosseyeux wrote:

At the beginning of the twentieth century, the situation in the hospitals of
Paris was, to say the least, not very splendid: there were complaints from all
quarters about the decrepitude of the buildings and the lack of adjustment to
the Pasteurian principles of hygiene; both the general public and the medical
profession were united in their request that the National Assistance Board
should make every effort to bring itself up to date with the advances of modern
science.[37]

Although it had been given top priority before 1870, salubrity in
hospitals had only just begun to be achieved. After the Pasteurian era, it
was relegated to second place and did not become a reality until much
later, largely during the interwar period.

Hydrotherapy and hygiene

It is well known that the first half of the nineteenth century was 'the golden age of alienism' (the study and treatment of mental illnesses) and methods of treatment based on hydrotherapy underwent considerable development during that period.[38] According to Michel Foucault, this use of hydrotherapy for the mentally ill dates back to the end of the seventeenth century.[39] In any case, since the royal decree of 18 December 1839, 'regulating public and private institutions dedicated to the treatment of the mentally ill', 'any person wishing to establish or manage a private institution for the treatment of the mentally ill [. . .] will have to provide evidence (Section II; article 22) that there will at all times be an adequate supply of good-quality water.'[40]

This particular medical use of baths was continued until well into the century, almost until the introduction onto the 'market' of neuroleptics and other 'tranquillizers.' In 1937, Dr Achard related his student memories:

At the beginning of my time as a houseman at Bicêtre, I saw the courtyard where disturbed patients were kept; narrow, dungeon-like closets opened on to it and inside them were bathtubs filled with cold water into which very agitated patients were plunged.[41]

However, the medical and therapeutic uses of water were not restricted to the mentally ill. Indeed, throughout the nineteenth century, baths, showers and steam rooms were used in the treatment of a number of more or less serious diseases in areas that were to develop into such specialisms as dermatology, venereology, rheumatology, psychosomatic medicine, infectious pathology, etc. In 1895, Bouchut and Després, both professors of medicine in Paris, listed thirty-three types of bath.[42] Baths for therapeutic and hygienic uses 'can be categorized, according to whether the body is fully or partially immersed, into complete baths or partial baths, which are either hip-baths or sitz baths, foot baths or hand baths, etc.'.[43]

Baths were further divided into single or composite liquid baths. The first kind were taken in ordinary water, either still or running; the second, known as composite baths during the Second Empire and by 1900 as 'medicinal baths' and 'mineral water baths' were taken in water to which animal, vegetable or mineral substances had been added. The baths to which mineral substances were added included acid baths (very rarely recommended), alkaline baths, used for rheumatic diseases, anaemia and certain skin complaints; artificial sulphur baths; iodine and chlorine baths; arsenic and mercury baths, etc. Those containing vegetable substances included emollient baths (with bran, starch, linseed, etc.), narcotic baths and aromatic, anti-spasmodic baths (containing lime, valerian, etc.). Complete or, more frequently, partial mustard (or poultice) baths were also prescribed. Finally, for children, baths of wine or wine lees were recommended as tonics.

Animal substances were used to make milk and whey baths for the treatment of rickets, chlorosis and neuropathy, blood baths and, more frequently, gelatine baths. Vapour baths, which were imported from Russia and Turkey in the 1820s, were also used, both in the form of hot-air baths and steam baths. They were given to patients in enclosed spaces – in bed, bedrooms, steam chests, bathtubs – and were used in the treatment of, among other things, rheumatism, skin problems and susceptibility to chills. The final category was solid or semi-liquid baths: grape marc baths were said to be 'a stimulant or a sedative, depending on the duration' and were recommended for rheumatoid arthritis; other baths of this kind were dung baths, tripe baths, mineral mud baths and sand baths.[44] Tripe baths were prepared 'with water in which by-products from horned animals have been cooked and which can be seen as a solution of gelatine mixed with a bit of fat'.[45]

Finally, baths were distinguished by temperature. A 'thermometric scale' consisting of six categories, was drawn up in the middle of the nineteenth century by Rostan:[46]

very cold	:	0° to 10°
cold	:	10° to 15°
cool	:	15° to 20°
moderate	:	20° to 25°
hot	:	25° to 30°
very hot	:	30° to 35° or 36°.

Of course these therapeutic uses varied considerably according to the illness, the patient and the medical theories of the doctor carrying out the treatment! However, one thing at least is well established: until the late nineteenth century, baths had a solid reputation with doctors, and not only the then novel hygienic baths, but more especially the therapeutic bath of antiquity. Ambroise Tardieu took pleasure in recognizing this in 1862: 'From the point of view of public health, the benefits to be gained from the popularization of baths are indisputable.'[47] And Michel Lévy, in voicing his approval, quoted Montaigne: 'In general, I consider bathing a healthy practice, and believe that our slight infirmities are caused by the abandoning of this habit.'[48].

With his position strengthened by this reference, Michel Lévy goes on to specify the benefits to be gained from hygienic baths, which were not at that time in widespread use, although the presence of the doctor in the hospital may have helped to diffuse their use.

The detailed examination that we have conducted of hygiene as it relates to the mouth, hair, faeces and urine, etc., leaves space here only for discussion of the modifications that can be made to cutaneous excretions and the surface that gives rise to them. The essential factor is the use of water at different temperatures in the form of complete or partial baths, ablution, lotion, affusion, sometimes with the addition of soap. . . . Water is the modifier *par excellence* of the excretory surfaces and the most efficient agent for disposing of their excretions and keeping them active and vigorous. Water can used beneficially on the skin, in the mouth and nose, the oculopalpebral mucuous membranes and the auditory meatus, which all give rise to a great deal of secretion. When injected into the rectum it stimulates the movement of the bowels; when applied in lotions to the genital organs it cleanses them of the residues left by their abundant secretions and reduces the risk of fatal contagion. The influence of water is not restricted to the general and immediate action of bath water on the epidermis; it spreads throughout the whole body, changes the rhythm of all its functions and returns them to a state of harmoniousness.[49]

For lack of antibiotics, sulphonamides and antipyretic drugs, baths were still seen in the nineteenth century as a general medication, if not as a universal remedy; this is the reason for their great diversity. The well-to-do classes often took them at home, while the working classes and paupers used shower baths and the public baths in almshouses. Hot baths 'stimulate perspiration and lead to general excitation, which is soon followed by lassitude, the extent of which increases as the temperature of the bath rises.'[50] Described as 'stimulating and revulsive', they were rarely used at the end of the nineteenth century, when they were prescribed for rheumatism, paralysis 'or when it was a question of creating a strong revulsive action on the surface of the skin (fevers with collapse)'.[51]

On the other hand, it was very common, at least until the end of the last century, to prescribe cold baths in cases of fever, particularly of typhoid fever. The same was true of showers; at the Ecole de Sèvres, founded in 1881, one regulation, which was in force for a long time, insisted that showers should be administered only on medical prescription and by a nurse.[52] Moreover, some doctors declared that cold baths 'correct within a few weeks the manifestations of first degree rickets in children'.[53] Many doctors in the eighteenth and nineteenth centuries, including Tissot, Cullen, Bordeu, Pujol and Bégin[54] himself, recommended them and stated that they had used them successfully in the treatment of scrofula. And Michel Lévy also praised them in 1862:

The use [of baths] is indicated in all cases where there is a need to stimulate circulation in the skin, to fortify the skin and the muscular system, to dull excessive irritability of the nervous sytem, etc. Many people whose constitution was originally weak have obtained from cold baths a vigour which enables them to withstand the fatigue caused by walking upstairs, to face the heat and cold with impunity, etc. There is no more certain way of countering a tendency towards tonsillitis, head colds, opthalmia, hoarseness, bronchitis, muscular rheumatism, sciatica and facial or cranial neuralgia; it is also indicated for the treatment of people predisposed towards obesity, tuberculization, scrofula, white swellings, bone diseases and chlorosis.[55]

Patients and hospital equipment

Patients and hygiene in hospitals

Open to the poor, the needy and the misfits in society, were French hospitals in the nineteenth century really privileged places and institutions in which, in conjunction with a new concept of hygiene, a new relationship with water was forged? Were they one of the instruments in the conquest of water?

It is undeniable that mistrust of water, based on its power of submersion, the dangers of cold and chill and its significance as the final 'journey', put a brake on the diffusion of the bodily uses of water among the poorer members of society who constituted the clientele of hospitals. 'According to long-established notions, a certain degree of dirtiness was necessary for children to protect their bodies, particularly their heads.'[56] Indeed, 'the dirtier children are, the healthier they are'[57] (proverb from the province of Limousin), even though the cosmic and sacred connotations associated with washing were evidence of 'the importance attached to the cleanliness of children's linen and clothes.'[58] Secondly, hospitals were often perceived as an alien, even frightening environment by the population of the time. In 1911, Rémy de Gourmont wrote in this connection:

> Those who have not been thrown into jail or who have failed to starve to death on their straw mattresses are sent there to finish their miserable lives. The stretcher, the death rattles and the sputum of the public ward, experiments with iron and poison, the anatomical theatre and the bed of lime: this is what those who remain standing have in store for those who fall.
>
> Hospitals, as they are organised in modern societies, are prisons for the sick and laboratories for doctors. . . Patients are books that are opened up to satisfy the curiosity of medical students. In Paris, hospitals are the terror of the poor.[59]

At Minot (Côte d'Or), both water and hospitals were feared: That the simple notion of the bath is associated either with birth or death is shown by the comments made about old people who, having been taken into hospital and bathed on arrival, did not return: 'it was the big baths that killed her off'; 'she died in the big baths', as if to be bathed was indeed to act out one's death.[60]

Even today, the central position occupied by baths and showers is perceived as a power of coercion exercised by doctors and paramedical staff; the same applies to supervisors and assistants in old people's homes, particularly the one in Nanterre where 'this power of touching, manipulating and poring over someone else's body'[61] is perceived as a dehumanizing loss of identity that becomes the inmate's lot as soon as he or she crosses the threshold of such an institution.

The Nanterre home is no exception to this general rule. Admittance into the institution is accompanied by the cleansing bath. Cleanliness is claimed by the authorities to be a necessity and is used by residents as a means of contesting the power of the institution [and symbolises the] purifying and hygienic role of the home.

It crops up constantly in residents' conversation: 'If you don't have your shower, you get a month in prison.'[62]

It is undeniable that hospitals, like advertising, impose a common, even 'totalitarian', way of thinking. Moreover, they establish daily routines from which patients can escape only with difficulty. However, since patients in the nineteenth century left little trace of their stay in these institutions, it is very difficult to formulate any precise idea of their attitudes and reactions to the hygiene practices that they experienced in hospital.

Hospital equipment

Thanks to the survey of hospitals conducted in 1864,[63] we at least know for certain what hydraulic equipment there was in hospitals and, to a lesser extent, what hygiene practices and hydrotherapy treatments were applied in them.

One initial observation is essential, and that is that the French 'hospital stock' was at that time extremely old, even dilapidated, despite the renovation and rebuilding programmes that had been initiated between 1840 and 1860: eighty per cent of the institutions surveyed had been built before the nineteenth century.

Secondly, the poorhouses and combined poorhouses and hospitals, of which there were very many during the Second Empire, were less well equipped than hospitals proper: twenty-eight per cent of the 397 poorhouses and combined poorhouses/hospitals had bathing facilities, compared with fifty-eight per cent of the thirty-five 'hospitals'. This gap is accounted for by the treatment of the mentally ill and of several diseases, such as typhoid and skin afflictions, that were the specific responsibility of hospitals. Indeed, the use of baths, which was unusual in this period for reasons of cleanliness or hygiene, arose out of medical prescriptions, particularly in hospitals. Thus in Landrecies (Nord), the destitute inmates of the hospital were not entitled to use the baths unless they were sick;[64] similarly, in Durtal (Maine-et-Loire), they were not allowed into the baths, 'unless they were ill.'[65]

If the truth were to be told, the coffers of the poorhouses were hardly overflowing during this period. The case of the poorhouse in Digne (Basses-Alpes), even though extreme, is conclusive evidence of this. 'When the buildings of the poorhouse were extended a few years ago, space was set aside on the ground floor of the new buildings for the establishment of bathing facilities.'[66] Overall, about twenty per cent of the institutions surveyed had seen relative improvements in matters of hygiene between 1852 and 1864. Hospitals accounted for a much higher proportion of these institutions than the poorhouses, located mainly in the north and centre of the country. These improvements concerned for the most part the supply and/or drainage of water and bathing facilities (seventy-three per cent of cases), and, secondly, the installation of WCs (seventeen per cent of cases) and wash houses (ten per cent).

However, there were bathing facilities and bathing facilities. Some were complete, other merely 'skeletal'. They consisted in principle of several elements: baths, sweating-rooms and showers. Various techniques were used, including single baths, composite baths and steam baths. In fact, the small number of bath tubs in each institution were virtually all included in the baths department, which is further evidence of their use for medical purposes. The proportion of hospitals equipped with bathing facilities varied from region to region, but the average stabilized around fifty-three per cent.

Once again, hospitals were better endowed than poorhouses. The overall picture was of a north/south divide, with the north of France being better equipped than the South. This significant difference, which had already been noted at the end of the eighteenth and beginning of the nineteenth centuries with respect to literacy and the height of conscripts, should not, however, conceal very wide local variations, particularly in the south of France. This is an indication that the existence of bathing facilities was not a response to sanitary standards imposed from on high, but rather a reflection of different situations and of the widely differing financial resources available to the parishes whose responsibility it was to administer their own hospitals.

In addition to the differences due to the type of institution and the region in question, the facilities available in institutions reflected fairly accurately the position of the city in which they were located in the demographic and politico-administrative hierarchy of Second Empire France. The most well-endowed hospitals were situated in cities with an average of 37,000 inhabitants. The less well-equipped institutions were generally located in small towns or even villages with an average population of between 2,000 and 3,000. Secondly, in fifty-two per cent of cases, the best equipped hospital was located in the department capital and in forty-eight per cent of cases in the chief town of the administrative district.

If the institutions included in the survey were relatively well equipped, only 16.7 per cent had showers, 22.1 per cent steam baths and 11.4 per cent sweating-rooms. In this respect, the contrast was not between north and south, but between eastern France, where the institutions were well equipped, and western France, where facilities were less than adequate. And, once again, hospitals seemed to be better provided for than poorhouses: thirty-seven per cent of hospitals had sweating-rooms, compared with eleven per cent of poorhouses, while fifty-seven per cent of hospitals had steam baths, compared with only nineteen per cent of poorhouses.

In 1864, steam baths and baths were the most widely used items of hydraulic equipment in hospitals, unlike showers and sweating rooms, which did not come into common use until later. Secondly, the difference between north and south France existed only in the case of baths, which is an indication of a higher level of medicalization, which had already been established for almost a century; in the case of other devices, the difference was between the east and west of the country, which may have been due to the new split which began to develop with the advance of industrial capitalism during the Second Empire.

The survey shows that the use of 'single baths' was twice as widespread as that of 'composite baths', which were no doubt more expensive. The regional distribution of these two types of baths does not reveal any major differences, except for a lower rate of use in the southwest. Under these circumstances, the regional inequalities in the provision of baths was not decisive, since more extensive use in the poorly endowed regions (such as central France) partially offset their lower level of equipment. On the other hand, the East, which was the region with the greatest number of baths – two thirds of the institutions were equipped with them – only made moderate use of them: thirty-eight per cent of institutions said that they administered only 'single baths.' However, in four of the six regions surveyed, there was a link in the majority of cases between level of provision and rate of use.

Since they were institutions for treating if not healing the sick, hospitals made more frequent use of 'single' and 'composite' baths than did poorhouses. Finally, the south of France turned out to be little concerned with making use of the baths that some poorhouses, at least, did possess. In other words, hospitals often made maximum use of their equipment, whereas 13.4 per cent of poorhouses did not use them at all, even when they had them at their disposal.

Thus the use of different types of baths, which varied greatly from region to another and from one town to another, revealed the inmates' fear of full baths, unless they were leaving the institution for the last time – which we do not know.

In fact, neither the hospitals nor the poorhouses were seeking at the time to spread the use of water or of baths, since sixty per cent of these institutions did not give the poor and needy of the parish free entry to their bathing facilities. Once again, however, there were significant regional differences. In the east, twenty-eight per cent of the institutions offered free access to the poor and needy, compared with nine per cent in the Southwest and an average of about twenty per cent for the rest of the country.[67]

However, with the spread of municipal shower-baths during the Second Empire, the poor and needy were admitted to institutions other than hospitals. Thus at Revel (Haute-Garonne) 'the relief committee will pay the costs' of the poor who 'take baths without charge in a particular establishment.'[68] However, the number of poor people asking for free baths at this time must have been very low, since full baths were repugnant to most people, particularly the working classes. Only one response in the 1864 survey provides evidence of this lack of interest. At Billom (Puy-de-Dôme), 'there has never been any request in this connection.'[69]

Under these circumstances, neither the provision nor the use of bathing facilities in hospitals and poorhouses is a reflection of daily life during the Second Empire. To the extent that such institutions were perceived as a foreign, even hostile environment and established links between certain uses of water and patients' diseases, they forced hygienic and therapeutic practices, some of which appear to us today as self-evident, even normal, and some of which are completely out-of-date, out of the everyday world of the healthy and into the world of disease.

The hospitals of this period, which were the favoured places for medicine and, to a lesser extent, for hygiene, and, particularly, the poorhouses, contributed very little to the diffusion outside of institutions of hydraulic equipment, of hydrotherapy and other uses of water; even though, in certain respects and only very distantly, they foreshadowed their evolution, they tended to repel their largely working-class clientele.

The doctor and his work

Between the Second Empire and our own age, considerable changes took place in hydraulic infrastructures and sanitary equipment. They proliferated to a point where they became extremely commonplace; and to a certain extent, social practices were modified in accordance with these material changes. The medical profession was a witness to and also one of the principal actors in this profound change.

In 1977, in the twilight of their lives, thirty-seven doctors – thirty men and four women[70] – in Paris and the surrounding region replied to a long questionnaire[71] and gave their judgement on the task that had been accomplished. Twenty-eight per cent were the sons or daughters of doctors and had been brought up by their mothers (53 per cent), who were present in the home in 79 per cent of cases, or by their mother and father equally (45 per cent). They had acquired typically urban hygiene practices which were virtually unknown among the working classes, particularly in rural areas. They were concerned mainly with the daily and general weekly toilet. To be more precise, they involved the cleanliness of the face, the brushing of teeth and the washing of hands before meals. The doctors made little mention in their replies of hair care, the cleaning of ears or intimate hygiene.

More generally, 75 per cent of the doctors who were questioned considered teaching about hygiene and cleanliness within the family to be important. On the other hand, most of them considered the instruction that they had received to be 'unimportant'. Thus lengthy studies – which at that time were not the norm – and their training in the Faculty of Medecine had not put them in a category apart. Like other people, they had learnt as children by copying and identifying with their parents.

However, this early education had a lasting influence. Habits acquired in childhood hardly changed throughout a lifetime, at least not to judge from their frequency before 1914, about 1930 and in 1976. Thus they washed their hands, ears and teeth every day, and even several times a day in the case of the teeth and hands. They also shaved every day, with one exception. Most of them washed their hair, and a minority their feet, once a week. Finally, personal hygiene was a daily habit for the women doctors, but for 18 per cent of their male colleagues it was only a weekly practice. Thus with the exception of foot washing, which increased in frequency between 1900 and 1976, washing habits among doctors in the Paris region remained virtually unchanged.

It is true that they benefited from a degree of comfort that was unusual for the period between 1900 and 1930, but which by 1976 had become much more widespread. They were born into well-to-do families and enjoyed a social status which required a good standard of cleanliness, if not of general hygiene, that was reinforced by Pasteurism; in general, the doctors had also enjoyed a high level of sanitary equipment since childhood. Most of them had always had running water at their disposition and, after 1920, domestic hot water. Only a few had had experience of water drawn from wells, springs, fountains or water tanks, generally before 1924. Three quarters of doctors had always had a washbasin, a bath and a bidet, and half of them a shower. Less than a third had used a jug and bowl for washing, which were widespread in the nineteenth century. Only 18 per cent had washed themselves in the kitchen sink and virtually none had used the buckets that used to be so common.

As early as 1900–10, virtually all of them were using bars of soap, toothpaste, shampoo and toilet water. The same was true of accessories such as toothbrushes, towels, face flannels and sponges. Finally, and very unusually for the period, 88 per cent had always had a bathroom at their disposal, or, at worst, a room set aside for washing, such as a toilet or a shower room.

The doctors were not much affected by the fact that they had to pay for water, to the extent that they found the price justified for reasons of hygiene, and had generally preferred town water for more than half a century. However, when it came to drinking water, they tended from the 1960s onwards to choose mineral water for reasons of health or taste.

Aware that they had received excellent health education, they were determined to pass it on to their children, and all of them said they had taken this task upon themselves. Finally, they were optimistic about their own hygiene practices (66 per cent) and those of their children (59 per cent), less so (50 per cent) about those of their grandchildren.

Secondly, these doctors had been determined during their professional lives (that is, between 1930 and about 1970) to implement principles of hygiene, particularly the 'elementary' rules, both among patients in their surgeries and in patients admitted to hospital. Half of the advice that they had given related in fact to frequent and regular washing. Other advice was concerned mainly with oral and dental hygiene, cleaning of nails and eye care. Finally, a quarter of the advice was aimed at reducing the consumption of drugs such as alcohol, tobacco and tranquillizers.

However, they were rather divided on the real effectiveness of their advice. The first reason for this was that, according to their own statements, patients changed their clothes and washed before going to the surgery or limited themselves to washing the part of the body to be subjected to medical scrutiny. The second was that many sick people were superstitious and that 'sick people consult their doctor in order to have their own prejudices confirmed.'[72] The final reason was that, with respect to bodily hygiene, doctors noticed only 'inertia and indifference' and that 'French people are naturally dirty.'[73]

Yet one of them, less pessimistic than the others, noted the 'normal submissiveness of indifferent people'[74] when he inundated them with advice. He made a distinction between his private patients, whose 'decorousness' appeared to please him and of whom he noted that they constituted an 'urban clientele of a satisfying social level' and his hospital patients who, in his opinion, included 'a number of distressing cases'. Another doctor made a distinction not between his hospital and private patients but between the town and the country: 'The situation in the town is fine, but standards of cleanliness in the country (in 1940) are much lower.'[75] Yet another recorded 'very real changes that have taken place in the past forty years. "Smelly" patients, or those wearing dirty underwear, have become a rarity in 1976.'[76] Finally, several doctors agreed on one point: there had indeed been change, but it was a recent phenomenon. A urologist and gynaecologist, having noted that it was not unusual 'in about 1936–1940 to admit rather dirty patients and send them for a bath before examining them', concluded: 'For the past ten years it has not been necessary to give advice of this kind [on personal hygiene]. There has been a definite change.'[77] However, few members of the medical professional were as optimistic as this doctor, aged 80: 'After forty years of general practice in France, I have an excellent knowledge of the subject and can assure you that hygiene in France is perfectly satisfactory.'[78]

In contrast, the image of hospitals has worsened. In about 1930, the public hospitals in Paris were considered to be models of sanitation and hygiene and were generally equipped with the hydraulic equipment regarded as 'standard' today. Nevertheless, 'at Arpajon until 1958 water for washing infants had to be brought in basins.'[79] Nevertheless, around 1930, the hygiene practices used by staff towards patients, the changing of underwear and bedding and the cleanliness of the public wards left much to be desired, according to a majority of doctors.[80]

In 1976, however, only 58 per cent of the doctors questioned considered hygiene standards in public hospitals in Paris to be 'satisfactory'. This was certainly not because standards had fallen, but rather because, in the world outside, standards and practices had changed and the expectations of patients and doctors had risen.

The medical profession – at least in Paris – was one of the promoters of hygiene and cleanliness. Convinced of their usefulness and even of their necessity, doctors did not breathe a word about working-class practices or of the resistance that they encountered from patients. Only two out of forty alluded to the existence of 'superstitions' and the 'usual routine' of their patients. Thus doctors were indeed the model representatives of hygiene and cleanliness, committed to the order, social splendour and level of health care that they symbolized. Convinced that they embodied a sacrosanct model of health which was superior to all others, their habits and practices remained virtually unchanged and they instilled them into their children and grandchildren and attempted to teach them to their patients. Thus, although their knowledge asserted itself as a total certainty, reflecting a belief in 'progress', their observations on the ground only rarely reflected its diffusion throughout society as a whole. The conquest of water, to which the medical profession has aspired for two centuries, made little use of their services as intermediary in the consulting room or hospital.

6

Schools and the moulding of attitudes

By about 1900, virtually all parishes in France had a state primary school. During the Third Republic, primary schools had become the favoured route for the diffusion of knowledge as well as centres for training the citizens of the future. Doctors and primary school teachers were both motivated by faith in science and progress, particularly because of their effectiveness in improving the health and happiness of mankind.

In 1882, at the very moment that religious education was ousted from lay schools, Jules Ferry, by way of a 'symbolic exchange',[1] had hygiene included among the compulsory subjects in the primary school curriculum. In the same way as the teaching of morals, civic education and history, it was intended to create durable foundations for a truly 'republican' republic.

For the founders of these schools, the scrutiny of doctors and the introduction of 'scientific' hygiene was easily justified, to judge, for example, from this definition of hygiene proposed by Dr Armaingand, President of the Anti-Tuberculosis League: 'Hygiene is the science that teaches us the means of maintaining and improving our health, of avoiding disease and living as long as possible. Thus, after moral science, which teaches us our duties and our rights, hygiene is the most useful of all the sciences and nobody should be ignorant of its basic principles.'[2]

Since it had long been the responsiblity of primary school teachers to teach morals and the rules of civility,[3] it seemed logical to the political authorities of the time that they should give instruction in the principles and basic practices of hygiene. This is why schools, viewed as work spaces, and primary school teachers, seen as 'the heirs and representatives of modern civilisation', constituted two complementary channels for the diffusion of the proposed social and medical model. Both were responsible to a large extent for its cultural transmission.

Marianne's wishes

As a staunch republican and good patriot, it was Marianne's wish that her children should enjoy the benefits of healthy, clean, airy schools which would offer sound, useful advice and be well disposed towards science and progress. The motive behind this desire was twofold: firstly, to provide the basis for revenge against Germany and secondly to demonstrate the merits of lay schools to the supporters of private schools, which were put in the same category as 'obscurantism'.[4]

It was for these reasons that the objective of the circulars issued by the Ministry of Public Education between 1888 and 1890 was to create an ideal space designed to maintain and strengthen the health of children educated in state schools. It was for these reasons also that the site allocated to the school had to be 'central, well ventilated, with easy, safe access and well away from any noisy, unhealthy or dangerous buildings and at least 100 metres from cemeteries'.[5] Once this protected site had been marked out, the layout of the buildings had to be determined on the basis of certain variables, 'in accordance with the climate of the region, taking into account the hygienic conditions, aspect, configuration and dimensions of the site, of doors and windows open to the sky and, particularly, the distance from neighbouring buildings'.

The layout of the classrooms was governed by a multitude of very specific regulations. The rooms had to be rectangular, with an area calculated on the basis of one square metre per child and a floor to ceiling height of at least 4 metres in order to provide the volume of air per pupil considered necessary.

Similarly, heating and lighting were subject to specific regulations. Thus Dr Delvaille, in an article in the *Medical Gazette* of 1880, was concerned with the amount of light admitted into each classroom and the amount to which each child was exposed. 'When the classroom is insufficiently illuminated, the children's eyes become tired from straining to obtain more light than is being admitted. If the classroom is excessively illuminated, severe problems, even illnesses, might arise, which M. Bouchardat, professor of hygiene at the Faculty of Medicine in Paris, has already brought to the attention of readers of the *Scientific Review*.' Finally, he noted – without paying too much attention to it – that schools were generally laid out for right-handed children: 'If the light comes from behind or from the right-hand side, it throws a shadow which is distracting for children when they write. This problem can be avoided by having the light shine in only from the left, as in Parisian schools.'[6]

As part of the same general concern with hygiene and the prevention of disease, the regulations governing school furniture were very specific.

Thus the writing-tablets had to be no more than thirty centimetres from the pupils' eyes and 'at an angle of between fifteen to eighteen degrees, without ever being less than fifteen degrees'.[7] With the same end in mind, benches fitted with a back rest were strongly recommended, and the teacher was requested to pay attention to the posture of his pupils when writing, in order to prevent scoliosis, for example. Finally, in their concern to prevent myopia, the doctors who were consulted stressed the importance of specifying the size and colour of the printed characters used in schoolbooks.

This bodily 'discipline' was not intended solely for the classroom where children were working under the teacher's supervision. It also involved the 'school house', the playground and − of course! − the lavatories. Thus there had to be enough WCs and taps for the pupils to make good use of them, without wastage or dirtiness.

These concerns with good order and hygiene were complemented by a general preoccupation with decency and morality. Pupils had to keep their hands *on* their desks, and the doors of the lavatory cubicles had, in principle, to have a space at the top and another at the bottom. This careful surveillance was intended to discourage pupils from indulging in auto-erotic practices. This concern with physical and moral rectitude is reflected in the dictum of Dr Henri Napias, one of the most respected authorities on hygiene in schools in the France of that time: 'Upright handwriting, upright paper and an upright body'.[8]

However, regulations governing the physical layout, the furniture and educational objectives of schools were not the only preoccupations of hygienists in the years between 1880 and 1900. Around 1885 in particular, a problem which gave rise at the time to several enquiries was the subject of lively debate in certain sections of the medical profession. This was the question of fatigue among schoolchildren and, more especially, of the alternating of periods of work and rest in the course of the schoolday. A vigorous campaign was being conducted at the time against the 'damage' caused by 'overwork at school', a campaign which was echoed by the National Academy of Medicine in 1886.[9] Dr Javal, the rapporteur for the Committee on Hygiene in Schools, specified in physiological terms the rule that primary schoolteachers should observe: 'It is particularly important that periods of rest should be sufficient in number to ensure that cerebral fatigue never reaches the point where attention begins to wander and short enough not to stimulate the circulation to a point that makes it difficult to start work again.'[10]

Not content with laying down standards to govern school buildings and school life for a whole generation, committed hygienists were also intent on organizing a national system of medical inspections in schools, both state and private, which were intended to ascertain, among other things, whether schools were operating in accordance with the regulations laid down by the authorities. By 1880, such intrusions by medical science were certainly nothing new. As early as 1803, Chaptal, a doctor and chemist, had planned to facilitate the diffusion of smallpox vaccination by setting up a dispensary in the chief town of each administrative district, with schools and hospitals sharing the responsibility for distributing the vaccine. This initial preventive measure, which was contemporaneous with the success of Jenner's method of vaccination at the time of the Consulate and First Empire, was hardly implemented at all and was coupled with the duty placed upon headteachers to notify the authorities of any 'epidemic disease'; the fear of cholera after 1832 certainly played a part in this!

As early as 1836, an initial system of medical inspection in primary schools in Paris was established. In 1847 and 1848, the Royal Academy of Medicine conducted an inquiry into an epidemic of typhoid fever in schools and colleges;[11] the law of 1855 on infant schools provided a regulatory framework for medical inspections in such schools.

However, it was from 1879 onwards that school medical inspections were organized on a systematic basis. The plan was to entrust the medical care of primary school children, who were of particular importance for the future of a nation with a low birth rate, to the custodians of medical knowledge. To this end, a body of primary school medical inspectors was set up, appointments to which were left to local initiative. In 1886–7, thirty-five out of eighty-seven departments followed up this government initiative, but only twenty-two went so far as issuing a prefectoral decree on the matter. In fact, shortly before 1890, only eleven departments had some sort of medical inspection system. Only four, including Seine, had made available the money required to pay for a team of medical inspectors.[12] However, in six other departments in which the prefects had not issued a decree, eleven towns had set up a system of school medical inspection. In Lille, for example, a very detailed form was filled in each month by the medical inspector; it related both to hygiene conditions in each school and illnesses suffered by pupils.

The department of the Seine was the first to establish a general inspection system in primary schools. Set up in June 1879, it was reorganized several times between 1883 and 1887.[13] From this period onwards, it consisted of 166 districts, each with 15 to 20 classes. In principle, each school was inspected twice a month. Each time a school was visited, the medical inspector filled in a form providing information on the teeth, eyes, ears and general state of health of each pupil. This heralded a period in which, from one end of the country to the other, a system of medical inspection was set up which monitored hygiene conditions in schools and the health of schoolchildren.

However, this drive to medicalize school life did not stop at establishing a secular primary school system based on the hygienist model or organizing a system of medical inspection for young pupils. Attempts were also made to change the attitudes and habits of pupils by using schools, the established dispensary of knowledge, for the diffusion of certain notions of hygiene. In this way, 'the most useful of sciences' would banish superstitions and ill-founded beliefs and inculcate habits intended to maintain health and fertility and thus the might of the French nation. And this is indeed what happened, thanks to the insistence of Dr Paul Bert, a physiologist by profession and a respected adviser on projects relating to state education, and also to Jules Ferry, who in 1882 made hygiene a 'compulsory subject' in state primary schools while at the same time introducing instruction in 'the fundamental notions of science'.

Hygienists then started to write the textbooks required to train teachers and educate pupils in accordance with the new programmes that had been drawn up. Some of them delivered lectures on hygiene in grammar schools and colleges of education. Finally, they put to good use the opportunities offered to them by school life, including speech days, in order to put their beliefs into practice and repeat the same litany of 'advice on hygiene, diet and clothing'.[14]

Nevertheless, these committed hygienists were not the only ones concerned to spread the teaching of hygiene. The school regulations of the period recommended that teachers should see to it that their pupils arrived at school in a suitable state of cleanliness. And they invited them to conduct a daily 'cleanliness check' at the beginning of each class. Thus primary schoolteachers, like the family and medical profession, became the agents by which a new notion of hygiene was transmitted. They played a direct role with respect to their pupils, but also an indirect role with respect to parents, in that they were often town clerks. In this role, they focused attention on the dangers of tuberculosis, the damage caused by alcoholism and drew attention to the new health regulations.

Through the books that they made their pupils read, the passages that pupils were made to learn by heart, the lessons they gave, the dictations they selected and their physical appearance and behaviour, primary schoolteachers in the *belle époque* embarked upon the task of propagating notions of hygiene and incorporating them into their teaching.

This is reflected in the following dictation intended for pupils in the junior forms: 'Louise does not like cold water. This morning she thought she had washed herself because she gently passed the flannel across the end of her nose. Her face stayed dirty and her hands black. Her mother did not want to kiss her in that state.'[15]

Here is another example:

Nursery school children learning the habits of cleanliness, 1885. 'Cleanliness is the first rule of hygiene' (Mané et Pugnère, *40 lessons in hygiene and morals*, 1902.) (Photograph: © Roger-Viollet.)

Cleanliness. I know a nice, studious little boy; I like him very much. However, the poor child has an unfortunate shortcoming: he is dirty. His hands are always black. His textbooks and exercise books are covered with stains. He does not wash his face, neck or ears. His hair is untidy. Thus he has become an object of disgust for his friends.

Exercises. Why do people not like this little boy? How does he keep his textbooks and exercise books? Does he look after his hair? What is a studious child? Underline the plural nouns. Conjugate: I know my duty. I wash my hands. I wipe and polish the brasswork.[16]

Similarly, the essay titles set for the Certificate of Primary Studies frequently related to 'order and cleanliness', such as this one set in Charlieu (Loire) in 1892 : 'You have visited a well-kept house and you are describing it to one of your friends. You give him your thoughts on the advantages of cleanliness and order.' This is the crib, put forward in an educational journal in 1893:[17]

My dear Jeanne,
We have been comfortably settled in the country for a few days now. We have
already made a few visits in the neighbourhood. My mother's best friend,
Madame Bernard, has had some major repairs done to her house and, since she
wanted us to see what had been done, she showed us all round her pretty house.
Ah yes! pretty, and as charming as could be! There is no gilded plasterwork or
other costly ornamentation here; no, the only luxury in Madame Bertrand's
appartment is perfect order and exquisite cleanliness. One feels that the eye of
the mistress of the house has alighted on each subject, that her searching glance
has peered into the darkest nooks and crannies. There is no dust to be seen
anywhere; the furniture and wooden floors sparkle. And what taste and good
order in the bedrooms! There is a place for everything and everything is in its
place. The mantelpiece is simply decorated, but good taste has governed the
choice of objects and scrupulous maintenance has preserved them and makes
them stand out. On leaving this agreeable dwelling, I said to myself: 'It is very
easy to decorate a house without going to great expense. A great deal of care,
order and a little taste, that is all that is required.' On arriving home, I set to
work. I used a polish for which I had kept the formula to clean the wood of our
old furniture; I cleaned the paintwork and filled the vases on the mantelpiece
with roses and flowers from the fields. My dear mother laughed a little at all
this activity, but deep down seemed very happy with it. I have resolved to
effect a revolution in the kitchen; our old maid servant will grumble a little,
but she will just have to put up with it. . . .

   Dictation passages, grammar and reading books all helped, through a
process of mimesis, to spread the new notions of hygiene. In *Le Tour de
la France par deux enfants*, which was the 'bible' of several generations,
the young André remembers the advice given to them by Madame
Etienne: 'My children, she had said to them, everywhere you go, nobody
will know you. Take care therefore to keep yourselves clean and decent,
so that nobody will take you for beggars or vagabonds. However poor
one may be, one can always be clean. There is no shortage of water in
France. There is no excuse for dirtiness.'[18]
   In another reading book,[19] it was the air that was celebrated, whether
it was the air in the idealized village in which the story was set, or the
air breathed by the pupils in their beautiful classroom in the country.
One typical incident occurs during a trip to Paris when the teacher, the
hero of the book, takes his class on a visit to the Paris sewers and
recounts a version of the history of the system.
   Hygiene and morals were thus closely linked. Decency, cleanliness
and tidiness, in the home as well as the body or the appearance of school
exercise books, were supposed to maintain good health. Moreover, they
attracted respect, sympathy or even love from others. Thus it was in
books on moral instruction that most space was devoted to 'hygiene' and
related notions. For example, in the book by Louis Boyer which was 'the
moral creed of a whole generation',[20] advice on cleanliness abounds from

the first chapter onwards. The following passage has a paternally reproachful tone:

Why have you not washed the tip of your charming little nose this morning? Why have you been miserly with water for your forehead? And your ears? Upon my word, although the task is unpleasant, I shall take the liberty of inspecting your ears as well. Ah! you have not even thought of your poor little ears! But don't you know that it is absolutely hideous to see a pretty ear, indeed the prettiest ear in the world, when it is not as clean and pure as mother of pearl fresh from the billowing waves?[21]

Central to this metaphorical comparison is the notion of a purity both physical and moral. This notion appears again in a circular dated 15 July 1872 addressed to the primary schools of Paris by the director of primary education: 'Cleanliness is almost always an element in and an indication of moral attitudes.'

Thus by the end of the nineteenth century, when the secular school system was expanding rapidly, a whole code of manners and 'hygienic morality' had been developed. The motive underlying its inclusion in the school curriculum turned out to be twofold.

Firstly, it was intended to teach children to think of themselves as social beings, whose appearance should not cause offence to others, particularly adults, but on the contrary attract admiration through their attitudes and cleanliness. Secondly, it was intended to benefit children by showing them that their own health was an asset that they had to maintain and enhance by adopting patterns of behaviour that would reduce the risk of disease. Based on shame and fear, this instruction was designed to integrate children into the ideal society of secular and republican ideology by inculcating a notion of 'self-constraint that makes individuals feel that they are behaving in a particular way on their own initiative and that they are doing so in the interests of their own health or to preserve their own dignity'.[22]

Finally, all these statements conceal some significant omissions. The advice on bodily hygiene is limited to visible, decent parts of a child's body, mainly the hands and face. There is no mention of hygiene as it relates to the rest of the body, particularly the genital and anal areas. The self-censorship implicit in the teaching of hygiene is thus clearly revealed; embarrassing questions and subjects giving rise to anxiety were avoided. Attempts were made in several ways to control the relationship of children to their bodies,[23] through the arrangement of school furniture, the forbidding of certain actions, the individualization of defecation, the vigilance of the teacher both in the classroom and in the playground and the development of physical education.

The schoolteacher as model

Over and above the instruction that he gave in matters of hygiene, it was the primary schoolteacher himself – and he was generally well aware of this – who was the image of the perfect adult for his pupils, their parents and for the community at large. It was his task to act as the living model of this hygienic puritanism. Eminently healthy and virtuous, the possessor of allegedly superior and, at that time, newly acquired knowledge, he was the 'Mr France' – symbolic in both name and sex – devised by Chabaud.

It was upon him that converged not only the attentive gazes of his pupils, but also the curiosity of an entire population. His influence was presented as authoritative in the area where he taught. Thus: 'Unsatisfactory pupils are unknown there, and the young people for about ten miles around seem to be of a quite particular type. They appear to be subject to a mysterious influence: their speech is more measured, their actions more restrained and their behaviour more dignified, with an unusual degree of gravity. One has the feeling that these youths will make fine men. Indeed, their reputation is almost proverbial in the local area. When one wishes to praise someone, one says of him that he is a former pupil of Mr France; nothing more needs to be said.'[24] There is no question that lay primary schoolteachers had their heart set on being models for several generations of children, particularly in matters of hygiene. The answers supplied by about fifty of them in 1977, when their average age was 78, provide conclusive evidence of this.

They washed very frequently, certainly more frequently than most of their pupils! They washed their hands and ears and brushed their teeth every day, even several times a day. They washed their hair once a month. Personal hygiene was attended to once a week. And the frequency of these habits acquired in childhood remained consistent, before 1914 as well as in 1930 and 1977. There are two exceptions, however: washing of the feet, which was a weekly custom before 1914, is now a daily practice, as is personal hygiene, particularly for men.

Unlike doctors, however, primary schoolteachers, who came from a lower social class and were less financially well off, were far from always having modern equipment and comforts at their disposal. Thus until 1914, two-thirds of them did not have running water, and they had to wait until the 1930s to enjoy the pleasures of hot water, baths and showers in the home. Until then, they often had to fetch water from a well, a spring or a fountain and to make do with the transitional jug and basin, the kitchen sink or a British-style 'tub'.

Finally, they had a very definite impression – much more so than the doctors – that their toilet habits had advanced, as had those of their children and grandchildren (81 to 89 per cent of responses). Thus for these schoolteachers, many of whom had begun their careers in about 1925, the lack of modern sanitary equipment was no obstacle to their adoption of modern toilet habits. The corollary of this was that modernization, when it finally happened, did little to change the habits they had acquired.

Thus Chabaud's 'Mr France' was not a purely mythical creation. He represented a type which existed and which contributed to the spread of the hygienic puritanism on which the Third Republic was based and thanks to which the standard of health in France was raised.

Habits called into question

By about 1930, the teaching profession, generally without reference to demographic or health statistics, were aware that they had 'triumphed'. They attributed their victory to their fidelity to the 'philosophy of the founders' of the state primary school system and declared that 'they were increasingly absorbing the inspiration of Montaigne, who wanted to make his pupil into a "spry and vigorous" boy, and of Rousseau, who applied himself to the task of creating, in "Emile", "an intelligent athlete".'[25]

The evolution of habits

Even if they did not always wish to choose between the supporters of the military tradition (that of Amoros), those of the physiological doctrine and those of the so-called natural method, the schoolteachers confirmed that they were able 'impartially to establish two things': 'The first is that public opinion has now been won over to the cause of progress, has rid itself of its superficial prejudices and supports the "physical renaissance" which we are now experiencing. The second relates to the results observed: children in both town and country seem to us sturdier, with greater strength and vigour and a more confident air: it is certainly not rash to attribute these qualities to a rational and determined programme of physical education.'[26] In 1930 as in 1900, primary schoolteachers justified the teaching of basic hygiene practices by data of a scientific nature and gymnastics lessons because they were intended to 'facilitate and stimulate the normal working and progress of the major functions (respiration, circulation, articulation) and to improve the coordination of the nervous system'.[27] Some schools even organized the school day around the needs of hygiene.

The good work to which educators as well as school doctors and nurses were committed was always based on the teaching of hygienic practices and habits, often directly or indirectly related to running water, which by 1930 was present in virtually all primary schools, which, according to the Hygiene by Example Association, should ideally have combined lavatories and cloakrooms and even showers, as shown in the Museum of Pedagogy in Paris. Without wishing to become too preoccupied with care of the body, because 'there are hygienic refinements' which may lead to neglect of the soul, and anxious to preserve their pupils from any narcissism, primary schoolteachers were utterly convinced[28] that bodily cleanliness and purity of thought were closely linked, that 'to respect our bodies is to strengthen our sense of human dignity' and that the moralist Jules Payot was right to say that 'the use of soap and a toothbrush was a sign of civilization'. Their pedagogic method consisted of linking actions to words and of encouraging their young charges to implement regularly and at a very early age the 'rational hygiene' which they cherished to the point of devoting things other than *ex cathedra* lessons to it. In about 1930, a lady schoolteacher in the Vosges was able to observe of her school that the 'showers, which were initially received with scepticism and even mistrust, are now looked upon favourably by almost all the families. The Saturday shower has become a reward for the girls.'[29] Henceforth, the 'water installations' were put in a place of honour, in accordance with official instructions; the WCs were no longer ignored but used in accordance with hygiene norms; the communal washrooms were used daily by pupils and wasting water became a cause for punishment.

In comparison with the Second Empire,[30] the material and cultural situation had changed considerably in the years leading up to the Second World War, although in some regions the process of change was extremely slow.[31] A little over a century ago, out of a sample of twenty-three *lycées* in twenty-two departments, eleven (47.8 per cent) had no facilities for boarders to take full baths, although all had a water supply[32] and WCs. However, in thirteen *lycées* (56.5 per cent), the pupils were taken to baths in the town. The cleanliness of the footbath rooms left something to be desired in four of them. Finally, in some *lycées*, those in the Caen educational district, for example, full baths were given only in cases of illness.[33] And, in 1858, the headmaster of the imperial *lycée* of Coutances wrote to the minister:

What is most lacking at the moment in the *lycée*, from the material point of view, is a daily supply of water for all the domestic services. This is the reason for the inadequacy of the full bath and footbath facilities. It is because of this problem that we are now experiencing some difficulty in retaining domestic staff who, in addition to their normal duties, are forced to go quite a long way every day to fetch the water they need. And yet it is difficult to emphasize this point too strongly to the municipal authorities whose concern it is. They are engaged in a project which will bring a large reservoir of clean water to the highest point in the town, from where it will be an easy matter to supply the *lycée*. . . . We are still waiting, however, and for five years the *lycée* still will be in a difficult and virtually intolerable situation with regard to the supply of the water necessary for the various services.[34]

In 1868, at the secondary school in Fontenay-le-Comte, the situation was hardly any better, even though 'drinking water is supplied from an excellent well which never dries up and which is always fresh and pure.'[35] However, there were no bathing or even footbath facilities for the pupils and they had to go to the municipal shower baths which were expensive; moreover, the journeys to and fro posed supervision problems and there was always a risk of the pupils catching cold on the way back.

Thus, from one end of France to the other, and with local variations, young boys were getting into the habit of taking baths and showers, in winter and summer alike, in accordance with the recommendations of the committees on hygiene. However, in many schools, cleanliness was still aproached with some caution, and footbaths were rather infrequent – every two weeks in Lille!

In fact, when the boarding regulations[36] are examined and former boarders in *lycées*, now aged between 70 and 80, are questioned, one thing becomes clear immediately: early rising, the frequent absence of heating and the lack of enthusiasm on the part of children and adolescents for their daily ablutions meant that, despite the wishes and the moralizing of the school authorities, the regulations on bodily hygiene were frequently not applied and the amount of washing was very limited,[37] even after a weekly shower became, in theory, compulsory from 1914 onwards.[38]

Hygiene and cleanliness

This was clearly realized by the Committee on Hygiene in Schools, set up in 1882 by the Ministry of Public Education which, in 1884, deplored the situation and intended that it should be remedied promptly: 'It must be admitted first of all that, of all the civilized nations, ours is one of those which cares least about cleanliness. . . . The most superficial enquiry is sufficient to prove that, even among the well-to-do classes, strict bodily cleanliness does not always extend beyond the visible parts of the body. . . . Habits are even worse among rural populations. One has only to practise medicine in the country to know the terror that the recommendation of a bath inspires in most peasants.'[39] However, the numerous and detailed responses from about fifty primary schoolteachers, both men and women, clearly show that they followed government recommendations and considered questions of hygiene and cleanliness to be an important part of their job. These responses thus make it possible to trace the broad outlines of the evolution of the fittings in the schools in which they worked in the period between 1920 and 1950.

In their view, the situation did not become satisfactory until about 1950–5. Indeed by then, all schools had WCs, 91 per cent had washrooms, but only 24 per cent had showers. Thirty years earlier, schools had been far from fully equipped; before 1925, 43 per cent of schools had washrooms, 50 per cent had WCs and 10 per cent had showers. In fact, between 1940 and 1950–5, the situation seems to have been stable: 93 per cent of schools now had washrooms, 96 per cent had WCs and 21 per cent had showers.[40] Thus, as the minutes of the meetings of the Central Committee for School Buildings show, the provision of sanitary equipment in schools improved between 1920 and 1940, even in rural districts.[41]

Under these circumstances, and with only a few exceptions,[42] it became easier for many primary schoolteachers to implement the recommendations of the circulars issued between 1883 and 1887. According to the replies obtained from the survey carried out in 1977, pupils obeyed, at least in school, the rules of hygiene and cleanliness that their teachers had instilled in them and which they ensured were applied by means of rules and punishments.

Pupils used the washrooms at least once a day, and often more. Similarly, they used the WCs, usually twice a day (according to 75 per cent of the replies) during each break or before each class, or else 'in case of emergency'. The use of these two hygienic installations was an obligation which was duly verified by the second-generation teachers (those born about 1900): 92.17 per cent replied in the affirmative with respect to the use of the washrooms, and 100 per cent in the case of the WCs. Similarly, the teachers complied with the responsibility placed upon them to conduct a daily cleanliness check. In general, they declared themselves 'fairly satisfied' with their inspections as far as the appearance of the hands, nails and hair were concerned; two-thirds of them were satisfied with nails and hair, and three-quarters with the hands. However, while there were few complaints about the washing of hands (only 5.5 per cent of replies), the state of the hands and nails was more often considered unsatisfactory (13 per cent of replies).

When asked to assess the general appearance of their pupils rather than the results of the 'cleanliness check', all the teachers declared themselves less satisfied. The majority assessed as 'good' the dress (63.6 per cent), cleanliness (53.3 per cent) and smell (50 per cent) of their pupils.

Thus both teachers and doctors made a clear distinction between the cleanliness of the visible parts and that of the unseen parts of the bodies of their 'clients'. However, rather than dress (9 per cent) or cleanliness

(11 per cent), it was the smell given off by their pupils which inspired the most unfavourable comments from their teachers. Consequently, in 1925 and 1955, their teachers were of the opinion that the younger generations had acquired bodily hygiene habits in school; moreover, the dress, cleanliness and smell of the majority of them was considered 'good' or 'average' (in 80 to 90 per cent of cases). This observation would seem to support the teachers' assessment of the general cleanliness of their pupils and give the lie to the doctors' pessimistic assessment of the hygiene habits of their own descendants.

Since the children's state of hygiene was very variable, to what did the teachers attribute these differences? At least one reason for the variations was mentioned consistently: the standard of living of families. They judged that the higher the family's standard of living, the more the child was encouraged to adopt adequate hygienic habits. In making this judgement, they showed that they accepted the notion of social differentiation, and also that they believed in the crucial influence of technical progress on hygiene, at least in the case of their pupils, whereas they themselves hardly changed their habits throughout the whole of their lives. On the one hand, most of them attributed these differences to the age of the children, stating that as they got older children paid more attention to personal hygiene. On the other hand, a large majority (62.9 per cent) did not believe that the child's sex was a cause of the differences observed. Among the other causes mentioned, the teachers included the availability of a good water supply, the need to be clean, which varied from child to child, and the fastidiousness of the mothers, which influenced their children: one teacher believed that children whose mother stayed at home were cleaner, while another mentioned the children of immigrant workers who had inadequate sanitary installations and different hygiene practices.

Although a large majority considered their pupils' dress, cleanliness and smell to be good or satisfactory, most of the teachers judged the sanitary equipment in the parishes or districts in which they worked between 1925 and 1955 to be nonexistent or inadequate. It was considered nonexistent by 16.3 per cent, while 61.2 per cent considered it inadequate and only 22.4 per cent judged it to be satisfactory. These views either reflected a real situation, or else the teachers' expectations in this respect were so high that they could not be satisfied. To decide which was the case, it is useful to see how they justified their opinion, even though only a small majority replied to this question. Out of twenty-nine replies, only three made positive mention of real progress that had been made, as a result either of new buildings that led to an improvement in sanitary installations, of efforts made by local authorities in this respect or of the introduction of a proper water supply and the installation of WCs.

On the other hand, from the twenty-six replies that mentioned deficiencies in sanitary installations, two things are immediately apparent. The first is that many teachers were unhappy about the late arrival of running water, particularly in rural areas. One of them recalled that running water was not installed until 1935, while another stated that in 1941 only 20 per cent of his pupils had running water in their homes. Even in towns, it would seem that it sometimes took a long time for running water to be installed in houses. Thus in two large provincial cities, running water was not installed in many otherwise comfortable apartments until about 1930; similarly, in Paris, many apartment blocks in the 15th *arrondissement* did not have running water until about 1930.

The second thing that emerges from these replies is that the teachers attributed these inadequacies in sanitary facilities to the policies of local authorities. It is true that they did point out that small rural parishes often had inadequate resources; however they also observed that prior to 1945, parish authorities generally made little effort to provide communal facilities. Some of them also noted the difficulties they experienced in obtaining from their district authorities the facilities required in their schools, notably running water, washrooms and WCs; some of them also indicated that, despite their requests, the facilities that were provided proved to be inadequate for the number of pupils in the school. Finally, some of the teachers mentioned the dilapidated housing, both in certain urban districts – such as the 11th *arrondissement* in Paris – as well as in some small rural parishes; in particular, they noted the lack of effective heating and the absence of showers and electricity.

Thus analysis of their replies shows that their demands do not seem to have been excessively high: although many of them demanded running water, not one of them required a bathroom in each home. They enjoyed facilities that were undoubtedly better than the national average and they wished to see these superior facilities spread throughout the country and the various social classes. This is why they felt they had to stress what they considered as 'inadequacies'. According to this survey, advances in sanitary facilities would appear to have been very slow and unevenly distributed. Did this slowness and unevenness have repercussions on hygienic practices in different regions and social classes?

Firstly, there was no unanimity among the teachers as to the real impact of their monitoring of standards of hygiene among their pupils; few of them actually expressed their views; one thought that the habits acquired in class 'had an influence within the family', while another observed that the emulation encouraged in school 'did not really conceal reality'. However, while some teachers stated that there had been an improvement in hygiene standards as a result of the diffusion of washrooms, then bidets and finally baths, and also of the development of urbanization, not all of them attributed these advances in hygiene to better technical facilities. Thus one of them stressed that old people generally had good bodily hygiene, even though they lived in old housing without modern comforts; another observed that children were properly washed and clean, even in the 11th *arrondissement* in Paris where there were a lot of dilapidated apartment blocks. A third teacher commented upon the use of coppers or basins instead of bathtubs for washing young children in a rural parish, both of which were equally effective when neither running water nor a washroom was available. On the other hand, some of them mentioned the use of muncipal shower baths in Paris, a practice which made up for a lack of private comfort at home. However, a higher level of equipment did not necessarily lead to better hygiene: one teacher noted that dirty nails were as common in the 15th *arrondissement* in Paris as in the rural parish that he had just left; the same was true, in urban and rural districts alike, of personal hygiene, the washing of feet and the smell of pupils.

On the other hand, very few teachers described the relationship that they postulated between better hygiene and the place of children in the social hierarchy. They merely noted that the children of poor parents had to wash in a basin, copper or tub, whereas the children of well-to-do parents had proper bathrooms. However, some of them said that, unlike middle-class children, hygiene standards in the families of farmers and black-coated and manual workers were 'very dubious', particularly at Orsay and in the departments of Loiret and Loire-Atlantique. There were not very many who noted any significant difference in this respect between town and country. They restricted themselves to pointing out that facilities were better in towns, but only one inferred from this that standards of hygiene were higher.

Some of the teachers mentioned regional differences. Thus hygiene habits were 'very advanced in the North' and in the region of Le Havre, less so in Brittany and 'very rudimentary' in the Sarthe. Finally, mention must be made of the big Sunday wash, a custom that was widespread in all regions and which meant that children 'smelt better on Mondays than on Saturdays'.

Only one-third of them noted the existence of customs specific to the regions in which they had worked. The others made no mention of them, either because they did not notice them or perhaps because they did not exist. The only ones who gave any details of such customs did so because their work in Africa and New Caledonia brought them face to face with cultural conflicts. However, a minority of the teachers did mention a few recent customs which encouraged the diffusion of hygiene: sea bathing in a district close to Le Havre, visits to municipal swimming pools or shower baths for the children of city-dwellers, the devotion of a school nurse and of a social worker who ensured that children at school and in the home observed the rules of hygiene. In several cases, however, the existence of modern sanitary facilities did not encourage better hygiene, as witness the cases of peasants who raised rabbits in their baths and Parisians who stored their coal in the same place.

More than half of the teachers mentioned the existence of certain prejudices against bodily hygiene, although only a few gave any details of the nature of these prejudices. Some blamed the attitude of the municipal authorities 'who saw no point in installing showers in schools'. Others mentioned a belief in the protection afforded by lice. Thus, 'the mother of one pupil insisted that her daughter should keep her lice because they protected her against illness'; similarly, some parents in rural districts believed that lice afforded protection against impetigo. Other beliefs were also mentioned: a daily shower was said to cause anaemia; washing hair in cold weather was supposed to cause colds; it was feared that children would catch a chill if they took showers in school, even in a heated room, and that washing the feet would inflame children's skin. Finally, as with the lice, there was a belief that 'it is natural to have fleas, because animals have them'. The factor common to all these beliefs was their hostility to the implementation of modern hygiene practices for children. They seemed to be based on the feeling that dirt and parasites were not only part of 'normal' life, but also played a protective role.

The teachers took it upon themselves to fight these beliefs and to spread notions of hygiene. Some of them inspected heads and handkerchiefs; others would check the cleanliness of their pupils each Monday morning (40 per cent claimed to have done this regularly each week throughout their whole careers). Most of them also gave lessons in hygiene, although the amount of time dedicated to this varied; some spent an hour a week, others a quarter of an hour per day. All of them insisted on two points intended to make their teaching effective; the first was that they should themselves set an example, and the second that the lessons should be short, simple and repeated. Nevertheless, the notion of hygiene was very variable in scope; most of the teachers made a strong connection between physical hygiene and moral instruction. In this latter case, the definition of hygiene could be extended to include virtually all aspects of existence; thus, for one woman teacher, 'mental hygiene is rectitude; social hygiene is unity; moral hygiene is liberty; sporting hygiene is care of the body.'

In other cases, on the other hand, hygiene lessons were restricted to cleanliness: washing, the cleaning of clothes, linen and the home, disinfection, cleaning cesspits and lavatories. There was frequent recourse to the textbooks used in science lessons. With the aid of the diagrams, they taught their pupils about digestion, respiration, the circulation, the skeleton and musculature. Each chapter of the textbook finished with some advice on hygiene. Thus during the study of nutrition, the teacher stressed the need for a balanced diet and the conditions under which food should be kept; similarly, when he was explaining respiration, he would teach the mechanics of breathing and advise that dwellings should be well ventilated. And when he was teaching about the skeleton, he would stress the importance of correct posture when pupils were writing or eating in order to avoid curvature of the spine; finally, he would praise the benefits of gymnastics for general balance and, more especially, for the suppleness of the joints.

All this instruction led to practical advice which the teacher could see to be well founded. He appealed to the intelligence and reason of his pupils in order to encourage them voluntarily to adopt behaviour that was forced upon them at school. Some teachers went further than others and did not limit themselves to cleanliness inspections. Thus one of them gave out 'blunted matches' for cleaning nails with; another insisted that his pupils should bring with them to class 'a few toilet requisites and something to polish their shoes with'.

In order to encourage their pupils to adopt real bodily hygiene, some of them resorted to the use of punishments and rewards; their pupils were excluded from school if they were repeatedly infested with lice, and pictures or 'merit points' were awarded when hands and nails were clean. They also often appealed to the children's moral sense; they were taught to identify with their teacher or parents in matters of cleanliness, they were read stories illustrating the physical and social necessity to maintain the asset that their health represented and to be well integrated with the adult world and were given lessons on the damage done to individuals and their home life by alcoholism.

The vast majority of the teachers were committed to inculcating their pupils with the precepts and notions of hygiene to which they themselves attached great importance, as a result of their education, training and personal habits. Even though they said that they themselves, and their own children, had no need of this teaching, they made considerable efforts in their own classrooms to develop an understanding of hygiene. From this point of view, the improvements to sanitary facilities in schools (mainly between 1925 and 1995 according to the survey analysed here) probably helped them in their task. On the other hand, the slow and unequal spread of these facilities may have reduced the effectiveness of their teaching.

In any case, the answers to this survey reveal that state primary schools were doubly favoured places in their relationships with hygiene and water. Firstly, facilities were installed more rapidly there than among the population at large. Secondly, pupils were taught how to use the facilities and use of them was made compulsory. In this sense, the teachers' individual and social experience in matters of hygiene met and merged. As supporters of technical and moral progress, offering themselves as examples, they attempted to pass on to their pupils their knowledge and their faith in the 'virtues' of water and hygiene in general.

A parallel system of dissemination

It should be stressed, however, that state schools were not the only agencies charged with the task of implementing this 'cultural revolution' which established a new relationship with water.[43] Doctors, government officials and hygienists were all committed, with a sometimes disconcerting naivety, to putting cleanliness and hygiene to the service of public health. Georges Clemenceau himself, doctor and founder of a dispensary in Montmartre, showed himself to have considerable knowledge of social problems among the working classes. This is clearly reflected in seven of his editorials published during the summer of 1904 and which were devoted entirely to the problem of white lead.[44]

In the editorial dated 4 September 1904, he declared:

I have never denied the fact in the case of a *hypothetical* worker painting with his shirt sleeves rolled down like M. de Buffon in the books. There is no doubt that a skilled craftsman, with access to a full set of toilet equipment and who spent a part of his day handling the appropriate brushes, *might often* avoid lead poisoning. It is quite a different matter for painters who have to grapple with the conditions under which most of them have to work. In order to earn as much as possible they have to get through the work come what may and whatever the risk. What good is achieved by a pompous, bespectacled civil servant declaring that if these proletarians worked in black suits and white ties and kept a safe distance from their painting they would avoid all risk! Working practices mean that this is simply impossible.

The average working man hardly seems to have had the time or the money to engage in a fight for his own health. He was caught between the need to maintain his output, from which he could not escape for very long, and the 'health recommendations' issued by government, factory inspectors and specialists in occupational medicine. Nevertheless, this was the goal towards which a minority of trade unionists and doctors were striving.

From 1905 onwards, an annual conference on hygiene and safety at work was held. The theme of the 1907 conference was the teaching of hygiene among the working classes. At the conference, Dr René Martial gave his impressions of a visit to Berlin, where he had attended an international conference on hygiene among the working classes. He observed that advances in hygiene had been more rapid in Germany than in France and that magnificent public baths, 'veritable palaces', had been built in all large cities. Proclaiming the fundamental principle of the need for individual hygiene, he invited general secretaries of trade unions and the managers of cooperatives and workers' associations to spread the message in their own organizations. He advised that the cause could be furthered by the holding of more conferences and the publication of booklets.[45]

At the same conference and on the same day, Dr Madeuf stressed the need for a worker to sleep for at least eight hours and made 'individual hygiene' a lower priority than the reduction of the working week and mass education. Delegates of manual workers rejected the Viviani plan which would have provided for worker delegates to the factory inspectorate because their role would have been limited to notifying the inspectors of infringements of the law committed in factories and because these delegates were to be financially dependent on the managing directors of companies. As the Corporate Union of Mechanics noted 'these delegates will be either the accomplices or the victims of management.'[46]

Thus the care of their own bodies became a fundamental aim for manual workers before the First World War. To this end, various associations and trade unions which supported the introduction of cleanliness, hygiene and safety in the work place expressed their aims and demands. In the periodical *L'Hygiène ouvrière*,[47] comrade E. Briat published a paper that he had given to the Conference of the Social Alliance for Hygiene, although not without giving rise to a few reservations on the part of the editors.[48] The title of the paper, 'The teaching of hygiene on the shop floor', is very revealing. It raised the following question:

Is it possible that the factory floor could make a contribution to transforming the health of the nation, both through instruction in the precise reasons for the need for hygiene and the secrets of preserving energy, and through a process of education which will gradually create new habits?

According to the author of this article, the application of hygiene measures which

are gradually spreading to all workshops, factories and shops has had the effect of creating an increasingly clean and healthy working environment for both workers and employers to spend about ten hours a day in, and of encouraging them to make frequent use of the facilities for individual cleanliness put at their disposal. Moreover, the habits acquired in the workplace are taken home by workers: thus, by example at least, they have become, more or less consciously, health educators for their family and neighbours; as a result, the shop floor is contributing to advances in general hygiene. In particular, it must be recognized that instructional and educative benefits can be gained from the use of special tools and working methods and of working clothes, from the importance attached in specialized industries to the extensive provision of cloakrooms and washrooms, baths and shower baths and finally from the establishment of a medical service in factories.[49]

And 'comrade' E. Briant quoted several examples, taken exclusively from cities and large companies in Paris, Nantes, Dijon, Toulouse, Lille and Lyons, where 'hygiene in the workshop is making such a contribution to worker education that large factories, which used to be condemned, not without justification, as inhumane and murderous, are now becoming the initiators of laudable progress in general hygiene.'

He also mentioned, though not by name, 'a few factories, not constrained by any legislation, whose owners have been guided solely by a desire to improve hygiene conditions and where shower baths and canteens have been installed'. He concluded, not without a certain degree of caution: 'Although progress has been slow, it has nonetheless been real and the notion that workplaces have to be kept clean and tidy is continuing to be absorbed even by small industrialists and traders. . .'

The argument is clear: since environment, in this case that of the workplace, changes people's attitudes, the introduction of hygiene and cleanliness into factories could not fail to influence manual workers. Workers would then introduce the values and practices acquired at work into their families. Thus the analogy that existed at the beginning of the nineteenth century between production methods in factories and the rational organization of the home (with Mary Pattison's *Domestic Management* echoing Frederick W. Taylor's *Scientific Management*) is taken up by E. Briant in this paper, this time on behalf of hygiene and cleanliness. The rationalization of work, in particular the saving of time and effort, is here applied to bodily health. It is argued that, when implemented, it will lead to a higher standard of health as a result of the habits acquired at work with respect to hygiene and cleanliness and, implicitly, to increased productivity. In this way, one of the great ideas that had driven the founders of the *Annales d'hygiène publique et de médecine légale* and which they had expressed in their 'Manifesto' of 1829, was taken up by a current of reformist thought.

L'Hygiène ouvrière also organized the 'Workers' Association for Hygiene and Safety' which had its headquarters at the Paris Labour Exchange. For an annual charge of six francs, subscribers received the periodical and the brochures published by the Association. The Association's aim was to 'determine and formulate the demands of the working class with respect to questions of hygiene, safety and insurance and to draw the attention of workers to all these questions and to emphasize their importance by all the means in its power (conferences, brochures, posters, etc.).

The rationalist, even scientistic campaign conducted in the state education system, particularly in the primary schools, was supported by a vast array of popular, quasi-scientific publications which had begun to emerge at the end of the *ancien régime*.[50] On the eve of the First World War, the amount of literature available was enormous. Serial sheets, specialist publications, monthly journals and conferences on hygiene[51] proliferated and found an avid audience, eager to discover authenticated opinons and scientific knowledge on matters of health. Usually edited by doctors, these sheets, brochures, periodicals, guides, manuals and even encyclopaedias were intended to instruct the public in 'accident medicine', 'medicine in the home', or even 'hygiene in 25 lessons'.[52]

Before 1914, this immense and extremely diversified literature was not intended for the masses. It was aimed at elite groups, such as priests, aristocrats and middle-class families, as the titles sometimes indicate. Towards the end of the nineteenth century, however, the readership began to widen; this trend became even more marked after 1900, when a radical reformist policy secularized education and the hospitals, expanding still further the state education system and the diffusion of the rules of hygiene. Items on public health in the newspapers were read avidly, sometimes even obsessively, by readers resembling the 'Monsieur Panard' humorously described by Maupassant in a short study published in 1886.[53] In general, the tone of the literature changed and the press became more diversified. The journal *L'Eau* and *La Petite Encyclopédie de chimie industrielle* (Paris 1898)[54] were intended for specialists and politicians, health officials, pharmacists and mayors; the lectures on hygiene organized by the *Annales* were aimed at a public often educated to *baccalauréat* level; in contrast, the *Feuilles d'hygiène* and their *Bulletin documentaire* seem to be addressed to a wider readership and, in particular, to all those who had sat the certificate of primary education. In this way, a very real concern with health was gradually democratized.

This vast programme of education was based on moral values and medical advice – óften closely linked – which were and still are shared by many medical practicioners and teachers and a large part of the French people. According to its defenders, it was effective because it raised health standards in France; in the view of its detractors, it led to excessive standardization which contributed to the development of a police state.[55]

This enormous cultural change, initiated in the primary schools and diffused through popular literature and the popular and middle-class press, was the subject of intensive debate, in which uncompromising positions were adopted. It reflected a new relationship with water, that is with the body and with nature, a relationship which incorporated the concepts of purity, cleanliness and hygiene. Finally, it involved a marked preference for social reformism and economic development. Its lesser 'shortcomings' included paternalism, scientism and colonialism.[56] All three reflected a superiority complex on the part of Western man towards his 'natural inferiors': peasants, aboriginals, women,[57] children, workers and the sick.

PART III

The Effects of the Conquest:
The Case of France

In the nineteenth century, water, an essentially free gift from God or nature, became an industrial product manufactured by man. Henceforth, duly educated citizens were prepared to pay the price that was now attached to it. In this way, water itself made new conquests: it changed the landscape and laid siege to daily life. More or less rapidly, the various social classes began to use it in their homes and places of work. It eventually became an essential element in bodily care in both town and country. Caught at times between the British model (that of the Industrial Revolution, notably) and the US model, France, true to its destiny, embraced the spirit of the times. In the year of the bicentenary of the French Revolution, let us hear her story.

7

Water becomes an industrial product

At the time of the French Revolution, when the concern with clean water first began to be manifested, water production was an artisanal industry. It was to remain so for a long time. In his *Memoirs of a nonagenarian*, the Frenchman François-Yves Besnard wrote:

At that time, the use of water filters was unknown; new arrivals in Paris thus usually suffered from colic, diarrhoea, etc. I myself suffered a great deal. In order to make the water less dangerous, I hit upon the idea of placing two carafes in my room and leaving water to stand in one of them for twenty-four hours; at the end of this period, it was quite frequently the case that, when the river was in spate, two inches of mud had been deposited on the bottom of the carafe.[1]

At the same time, however, the first private company to produce and sell purified water was set up in France. Mercier, always well informed, exclaimed in his *Tableau de Paris*:

A company is being set up to sell us water from the Seine! The company is manufacturing some sort of liquid that it claims has been purified. What does this prove? That for three-quarters of the year the Seine is muddy. . . . Water from the Seine has to be purified in one's own home if one wishes to drink clean, healthy water. Twenty years ago we drank water without paying too much attention to it; however, since the family of gases and the race of acids and salts appeared on the horizon. . ., much thought has been given to the pronouncements of chemists. . . . We began by analysing water and we now think when we drink a glass of it, which our oblivious forebears did not do.[2]

In fact, as early as the second half of the sixteenth century, the quality of water was concerning politicians and scientists.[3] However, it was during the second half of the seventeenth century that the idea of purifying river water began to gain a hold among enlightened folk, both in Paris, with respect to the Seine, and in Nevers, for the 'silt-laden water of the Loire.'[4] For all that, the use of domestic water filters, which seemed to be fairly widespread among the middle classes, was far from being always effective and provided only small quantities of drinkable water. By the beginning of the nineteenth century, many middle-class people would have liked to have an unlimited supply of water. Achille Dufaud wrote to his father from England:

> We lack one capital thing, and that is a water supply in the home. This provides great comfort, greatly facilitates the running of the household and makes possible a marvellous standard of cleanliness; with a compression pump and some lead pipes you will be able to make good this lack, but this expense is necessary for comfortable living. . .[5]

Demand of this kind could not help but lead to the expansion, hesitant at first and then on a massive scale, of the industrial production of water, all the more so since it depended upon the 'London miracle'.

Like other large French or German cities of the same period, London had the great conduit in West Cheap, which was begun in 1285. At the end of the sixteenth century, there were twenty conduits, three of which gave their names to streets in modern London: Great Conduit Street, Lamb's Conduit Street and White's Conduit Street. In 1582, a Dutchman, Peter Morris, built the London Bridge pumping station. From that time onwards, the principle of granting a water supply to private properties was adopted. London was two centuries ahead of Paris and some other Western cities. At the beginning of the seventeenth century, the Lea River in Hertfordshire provided extremely pure water, due to the construction of an aqueduct, which was further improved in 1613. Throughout the seventeenth and eighteenth centuries, other private companies were set up, each of which operated in their own region within the framework of a market economy. In 1828, there were nine water supply companies, supplying 164,000 tenants in a city with about 200,000 houses and 1.5 million inhabitants. From this time onwards, the quality of the water supplied was the subject of all sorts of disputes (in the medieval sense) and great care.

A precursor: the Paris Water Company

Well before the first industrial revolution, therefore, this new perception of water and its uses, inspired by the British model, had created considerable potential needs, both in terms of quantity and of quality. Hitherto purely scientific projects were now to come face to face with the capitalist spirit of enterprise.

Once the proposals drawn up for Paris by the academician Deparcieux had been rejected,[6] supporters of steam engines and those of hydraulic machines came into direct conflict, with the former being represented by the Périer brothers and the Chevalier d'Auxiron and the latter by the architect Capron. On 27 August 1778 the first capitalist water supply company in France was set up. It was a limited partnership, with capital of 1,440,000 livres[7] divided into 1,200 shares each with a face value of 1,200 livres. On 31 August of the same year, the first shareholders' general meeting was held. On 8 August 1781, the first official test of the steam pump took place, an occasion which marked the success of the Périer brothers.

At that time, two machines were used to produce water for consumers in Paris. On 31 August 1784, the water thus supplied was described as 'very wholesome' by the Royal Society of Medicine. In the meantime, the company had increased its capital, and in 1785 it took over the rival company Vachette for the sum of 150,000 pounds. This gave it a virtual monopoly over the supply of water in Paris, particularly since it installed the Gros-Caillou machines in the same year.

The early years of the Paris Water Company were fraught with difficulties: the number of subscribers grew only slowly and competition from the water carriers remained very strong. Until 1785, the carriers obtained their water from the rival pumps of the Vachette brothers, whose water had a better reputation. Looking at the situation with a certain degree of realism, the Périer brothers decided to cut by half the price charged to water carriers for a hogshead of water; this reveals a great deal about the fixing of the price of water at this period! Without neglecting sales made through carriers, they decided to expand sales by subscription. To this end, the company offered to supply subscribers with a hogshead of water per day (238 litres) for the very favourable price of 50 livres per year, or half a livre per cubic metre. Those water carriers who delivered their wares in buckets carried on a shoulder harness sold their water for a fixed price per floor of an apartment block; excluding both tips and surcharges (per floor), their price was 2 sous for 30 litres of water, which was a little higher than the carriers who delivered water in barrels, but much more than the rate proposed for subscribers. Indeed, the price for water delivered in buckets was 3 to 4 livres per cubic metre, that is six to eight times higher. It is true that in the Paris of that time there were many sources of water for domestic use. The well-to-do could send their servants to fetch water from the river, from pumps, from springs or from fountains in Paris, Auteuil or Passy. Moreover, the greater distance people lived from the Seine, the greater was the tendency to draw water from wells.[8]

Finally, the average amount of water to be allowed per person was the subject of much discussion at the time. Opponents of the company, such as the Marquis de Mirabeau, fixed the norm at 10 litres, while members of the company decided on 20 to 25 litres. The debate was so heated that at the end of 1785, when Mirabeau's terms were made known, the value of the company's shares fell by 500 livres from 3,000 to 2,500 livres (approximately). Speculation was in fact rife. The value of the company's shares soared and fell back suddenly in reaction to attacks from its economic and political opponents. In 1787, the *Entreprise de l'Yvette* won a victory when the king authorized Defer's proposals which slashed the price of water.[9] Supported by the Government and the city authorities, the Paris Water Company fought back; however, it was forced in April 1788 to enter into an agreement with the city authorities in order to obtain enough customers to guarantee its profits. Faced with the reluctance of Parisians to use its services, it had to be nationalized; the company then came under the financial and political control of the Government. However, this had little effect on the company's business: in 1788, according to the report made to Necker by the Marquis de Gony, the Périer brothers supplied only 900 hogsheads (209.7 cubic metres) of water per day.

At the beginning of the nineteenth century, the situation changed, with the Chaillot and Gros-Caillou engines supplying two-thirds of the drinking water consumed in Paris (6,509 cubic metres per day out of a total of 8,801). Although the quality of the water was still the target of criticism, the Chaillot engineers had provided a practical solution to the city's water supply problems. The Périers were not only passionate supporters of steam pumps, they were also the forerunners of the domestic water supply. They offered an alternative to the water carriers which was, at the time, a daring innovation. In 1888, Belgrand himself recognized this:

The company's prudent, rational propsectus for 1781 proves that the Périer brothers were fifty years ahead of their contemporaries in France; it was only twenty-five years ago that their system was fully implemented in Paris.[10]

Moreover, their steam pumps outlived the engineers of Chaillot: some were in regular use until 1853, others until 1858.

There is no doubt that the Périers modelled their system on the work of British engineers, and the criticisms levelled at it were not all inspired by a concern with public health: lack of understanding and deeply ingrained hostility, together with competition from their rivals, acted as a brake for a long time on the expansion of their company. Nevertheless, thirty years after their initial steps to secure the right to supply water from the Seine by means of steam pumps, the city of Paris found itself relatively well supplied with water, as did Parisians themselves who benefited from a more abundant supply of water in the city's fountains.

The company set up by the Périer brothers in Paris at the end of the eighteenth century was thus the forerunner of the capitalist enterprises (both state capitalist and private companies) engendered by the trade in water and the need for a water supply system. Throughout the first half of the nineteenth century, there were many companies manufacturing and selling domestic water filters and lead and then cast-iron piping, as well as companies specializing in drilling artesian wells.[11] These companies would certainly be worthy of a detailed study which would place them in their local context and reveal the nature of their products, their prices and their spread. For lack of a quick summary, it will be necessary to conduct a case study, which will at least be sufficient to show the influence and wordly wisdom deployed by one large capitalist company, the Compagnie Générale des Eaux, French by origin but, from its inception, an international enterprise.

Quiet strength: the Compagnie Générale des Eaux (1853–1940)

Established by company deed as a limited liability company and authorized by an imperial decree dated 14 December 1853, the Compagnie Générale de Eaux was intended to fulfil the following aims: 'In consideration of the important contribution that the establishment of a company with the objective of providing a water supply in cities and of draining land could make to the adornment and salubrity of cities, as well as to agriculture and sanitation in rural areas, they [the persons appearing before the court] have resolved to put into practice this scheme, which will be of such benefit to the public.'[12] This is a succinct summary of the fundamental objective. The company was set up not to generate profits but to offer a service to society as a whole. Mobilizing public interest to its own advantage, it harnessed the two qualities traditionally attributed to water: making land fertile and imparting its purity, that is health, to the inhabitants of the most polluted places, in this case the cities. It should be pointed out that improving the health of peasants, who were in 1853 the largest social class in the country, does not seem to have been considered a worthy objective.

In this period of economic prosperity, the report of the board of directors dated 26 October 1853[13] set out the 'general considerations underlying the formation of the company': 'In the new era which we are now entering, be assured, gentlemen, that massive sums of money will be devoted to the installation of water supply systems, just as in the previous two periods, enormous sums were spent on railways, roads, waterways and shipping.' Nevertheless 'in the area of irrigation. . ., even though there are schemes as fine as the Saint-Germain and Orléans railway projects', the Compagnie Générale des Eaux proceeded with great caution, and declared that it would wait for alterations that it was requesting the Government to make 'in the legislation governing these sorts of matters'. In fact, the board of directors wanted to base the company's operations on 'simpler projects, easier to assess and implement' and stated its intention of starting with the supply of water to cities by providing 'assistance for municipal authorities in implementing schemes of fundamental importance to public health'.

From its very inception, the company received support and encouragement from the business community, as a list of its founders shows. These included a Rothschild, a Fould and a Lafitte and the Pereire brothers as well as Government representatives; a minister (Persigny), the half-brother of the Emperor (Morny) and, in addition, a large part of the (mainly imperial) nobility were also included among the company's first shareholders. In the technical sphere, the company enlisted the support of some excellent engineers, notably from the Highways Department. Finally, even before the company was set up in 1852, one of its future managers 'had on his own behalf put forward to the city of Paris proposals relating to the supply of water'; these proposals in fact constituted the origins of the company. Proud of announcing the establishment of a company that was a pioneer in this field, the first board of management stated to its shareholders: 'We shall be opening up a mine, the wealth of which has not yet been explored; as the first occupants of this mine, it will be our privilege to select and exploit the best seams.'

Indeed, at the level at which it pitched its operations from the very outset, the Compagnie Générale des Eaux was the first major capitalist enterprise in France[14] concerned to make a profit from operating a water supply system in towns and cities. In fact, and this was stressed by the board of management from 1853 onwards in order to reassure the water carriers, this type of enterprise was not really an innovation in the major industrial countries, notably in Britain and the USA, where such companies already existed. The bankers who took part in the establishment of the company were well aware of that. They went to great pains to gather accurate information, and were able to declare to their shareholders:

If we are to judge from the examples of England and America, water supply companies in towns and cities afford excellent prospects. We do not wish to draw your attention to the most profitable company of this type, New Rider in London, whose shares now have a return of 1,000%, but we have discovered that these companies are generally sound, particularly where they are not in competition with each other; now as you know, these sorts of concessions are in France a prerogative which do not allow of competition.

The application for shares was obviously awaited with impatience in 1852–3, the period when the company was drawing up its statutes and might have been able to command a significantly larger capital than that decided upon. Finally, only a few privileged people were actually in a position to apply for the 80,000 shares issued at a price of 250 francs each: within a few days, payment to Rothschild Brothers, bankers to the company, was complete. The beneficiaries were bankers, stockbrokers, men of property, *rentiers* and dealers. James de Rothschild was the largest single shareholder with 5,000 shares, while the nobility held more than a quarter of the issue. The annual rate of return on each share was 4 per cent.

The Lyons project

The first project on which the company embarked was in Lyons, where the question of a water supply system had been debated and studied for more than twenty years by the municipal authorities, engineers and industrial companies. The schedule of works had been drawn up with care, and expensive and detailed experiments on the quality of the water and methods of purification had been carried out before the Compagnie Générale des Eaux entered into an agreement with the city of Lyons. The works were carried out promptly and the supply of water began in 1856. The Compagnie Générale des Eaux was granted the concession for the public water supply, subject to some interesting conditions. The city of Lyons undertook to take delivery of a maximum of 10,000 cubic metres at a rate of 17 francs per cubic metre for a period of twenty years and at a rate of 15 francs for the next seventy-nine years.

In other words, the company had a guaranteed annual income of 161,500 francs for twenty years and 142,500 francs for a further seventy-nine years. Moreover, this annual income was likely to rise as a result of the highly probable increase (from 1853 onwards) in the city's consumption of water.

Secondly, the Lyons Gas Company, in order to have its concession renewed and as a result of the authorized cooperation between the two tendering companies, had agreed to pay the Compagnie Générale des Eaux an annual sum of 100,000 francs for ten years. Thus even before the company had begun to explore the industrial and domestic subscription markets, the company already had a guaranteed annual income of 261,500 francs. In 1853, the various companies bidding for the concession had estimated the subscription income from the major industrial companies in Lyons, particularly dyeworks, at 120,000 francs per year. Consequently, from its first year of operation, the company could count on a gross income of at least 381,500 francs, leaving aside any income received from domestic subscribers.

The Lyons project thus made one thing clear: the supplying of water to private homes was only a means of adding to revenue; in fact the

Compagnie Générale des Eaux negotiated as one strong body to another, and preferably with secure institutions. The investment required for the Lyons system was not enormous; total expenditure on filtration galleries, drain tanks, machinery, buildings, reservoirs, mains pipes and drains was six million francs. This figure even included the compensation paid by the Compagnie Générale des Eaux to a Lyons company in exchange for its surveys and plans, in addition to a very advantageous market for the purchase and laying of cast-iron pipes.[15] Operating costs, at 80,000 francs per year for the first few years, were also low. Guaranteed income was thus about 300,000 francs per annum. At worst, the net profits would have been 5 per cent of tied-up capital. In fact, from the outset, they reached 25 per cent, since the forecasts of demand proved to be accurate.[16]

The other markets

Encouraged by this success, the Compagnie Générale des Eaux negotiated similar agreements with four other local authorities, including Nantes. Similar studies were carried out in fourteen other towns and cities, including Paris, which represented an enormous market. In the case of Nantes, the water supply system was the subject of investigation over a period of several years by Government engineers resident there. The proposal submitted was approved by the Civil Engineering Committee and was declared to be in the public interest by a decree of December 1853. The contract was signed in 1854.[17]

The city of Nantes had desperate need of drinking water, which its fountains failed to supply in adequate quantities. It was a rich, commercial city, keen to have fresh water for its shipping and industries and was an ideal client for the Compagnie Générale des Eaux. It could afford to purchase water at 0.80 francs per cubic metre, a price that would be profitable for the water company, but which was considerably less than the excessive 3.56 francs that the city had been paying until then. Thanks to the Compagnie Générale des Eaux, therefore, the price of water to consumers fell considerably. For all that, the (net) profits made by the Company were not exactly philanthropic: they totalled 20 per cent per annum from the start of operations.

However, the Lyons and Nantes schemes were only an encouraging start. The big prize was Paris. From its inception, the Company had expressed its interest in providing a water supply system for the suburbs of Paris. Proceeding in a calculated way, it gradually took over various local companies; it was in this way that the present water supply system in the Paris suburbs came into being.

In 1860, however, these same suburbs located on the periphery of the capital were annexed to the city. Exploiting this new situation, the

Compagnie Générale des Eaux entered into a fifty-year agreement with the prefecture of the Seine and the city of Paris. The Company transferred ownership of all its land, buildings and machinery to the city authorities. In exchange, the city authorities gave the Company control of the water supply to the new districts of Paris. The dividing line was thus clearly defined: buildings, maintenance and technical services were the responsibility of the city authorities, while the Company was responsible for commercial services. In other words, the city owned and managed the infrastructure and raised taxes from the population to cover its costs. On the other hand, the profitable part of the enterprise was given over to a private company.

Between 1860 and 1914, the Compagnie Générale des Eaux developed its network in the Paris region. In 1867, the city of Paris authorized it to negotiate with those districts in the department of the Seine that the city was no longer going to supply itself. In 1869, moreover, an agreement was concluded that gave the Company a free rein in all the suburbs of Paris. This gave the Company access to the whole of the Paris conglomeration. The Franco-Prussian war of 1870 even gave it an opportunity to strengthen its position by supplying the new forts that encircled Paris. A contract subject to periodic renewal henceforth tied Paris and the suburbs to the Company.

The Company's initial expansion can thus be easily explained. Long-term contracts guaranteed its income in large cities. Because it had speculated on the potentially very profitable market offered by a water-devouring economy, the prudent, far-sighted Compagnie Générale des Eaux made an enormous contribution in its early years to the conquest of water.

Not content with supplying large cities, the Company introduced a regional policy from 1860 onwards. In 1860, the annexation of the earldom of Nice brought a new jewel to its crown. Four years later, it concluded a concession agreement with the city of Nice, which was then extended to other towns in the region such as Villefranche-sur-Mer, Monaco, Menton and Antibes.

Generally speaking, the Company pursued a consistent policy of investing in prosperous regions and cities that were able to appreciate and pay for its services. Between 1880 and 1900, it became established in the Rouen area and in the resorts on the coast of Normandy. It embarked on projects in three large cities in Brittany, including Rennes, and did not neglect to include in its plans the resorts on what is now known as the Emerald Coast. In the North and East, it chose Arras, Boulogne-sur-Mer and Lens and the surrounding mining district. Between 1920 and 1940, it stuck to its earlier strategy and developed its facilities in the same regions. The *belle époque* was thus an essential phase in the growth of the company and, more generally, in the conquest of water.

With the exception of the area behind Nice, the Compagnie Générale des Eaux had concerned itself essentially with the second objective that it had set itself at its foundation, which was to make available to individuals and industry water of good quality which would meet the incessantly growing needs of a water-devouring economy[18] at a reasonable market price.[19] In short, it made water into a new mass consumption product in an industrial society.

The price of water

Water is like health: it does not have a price. In other words, the price of water is determined in a way that is contingent upon prevailing economic circumstances, in a manner more authoritarian than objective and in accordance with location and need. If inflation is disregarded and services rendered are taken into account (supply costs, compliance with quality standards, maintenance), three observations can be made:

the price of water increases only slowly in the long term;
the price gives rise to profits that can be immediately reinvested in other
   economic or social activities or in self-financing;
it makes it possible to experiment with different management systems
   and to increase their efficiency.

In the British system based on a public service paid for out of taxation, the tax on water (water rate) is linked to the value of the individual property. For this reason, there are no water meters in Great Britain. In France and other francophone countries, on the other hand, water is a market commodity. Consumers receive a water bill proportional to their consumption.

It is thus easy to understand the coexistence of two different types of water industry. In the British system, long-established private companies (set up in London as early as the end of the sixteenth century) played a leading role until 1847 but then declined in importance with the expansion of 'water and gas socialism' (1845–7). This type of management structure was a product of municipal socialism, which was then in its heyday in both France and Germany, but particularly so in Great Britain, which was its real birthplace. In all three countries, the municipalization of public services was then at its height. Even today, Great Britain bears the stamp of this heritage. Twenty-eight small private companies control only 20 per cent of total consumption.

In France, on the other hand, five large private companies are responsible for the water supply in 45 per cent of local authority districts, which account for 57 per cent of the population and 18 per cent of total water consumption. The largest company dates from 1853 and today controls 51 per cent of the private water market. It has used the regular income from the sale of water (it sells water at the price the consumer can pay) to build up over the past thirty years a veritable industrial empire with interests in sanitation, domestic waste disposal, public works, housing, heating systems, fire-fighting, undertaking and street furniture. In conjunction with its junior partner (the Lyonnaise des Eaux, founded in 1881, which controls 25 per cent of the market), it has conquered certain market segments in Europe (France, Belgium, Italy), the New World (Canada, the United States, Argentina) and the Far East (Indonesia, Japan).

For lack of a study covering the entire Western world, a gap which is felt more today than ever before, we shall once again examine the case of France in the nineteenth century. At this time, neither of the current systems was fully developed in France. The water industry was not yet totally private, nor still completely public, and water meters were still a rarity. In view of the complexity incessantly engendered by the interlocking of time and place, it is advisable to leave international comparisons until a later date and concentrate instead on regional studies, in the sense that France is a region of the world as a whole, particularly of the Western world.

By 1930, the majority of French households were supplied with water; it was produced by numerous private and public companies with different management methods. Since the large private companies were mainly interested in the more profitable sectors, those districts which considered themselves poorly served benefited from 1902 onwards from state subsidies. The supply network then expanded more rapidly, particularly since the profits to be made were generally not inconsiderable, even when water was sold by the municipality to its inhabitants. Between 1900 and 1904, very large companies such as the Compagnie Générale des Eaux and the Lyonnaise[20] existed side by side with a multiplicity of local and regional companies. Because of the law of concentration, these small companies disappeared, particularly after 1940, while the market share of the two largest companies gradually increased.

Management methods

Small and large companies alike managed the water supply in accordance with two methods, one direct, the other indirect. Both had been developed in the nineteenth century and still exist today.[21] In the indirect method, the administration and management of the supply network is contracted out to a third party, either with or without a financial interest in the business. If the third party has no financial stake in the operation, an agreed price is paid for the management of the supply network. The price of water to local inhabitants is fixed by the local authority, which retains any profits that are made.

The third party may, on the other hand, have a financial stake in the operation, either through a bonus scheme or a profit-sharing system. Moreover, the operating company may have been granted either a concession or a lease. In the concession system, developed between 1800 and 1850, the concession holder enters into an agreement with a district (represented by its mayor) to administer the public water supply, and charges users a fee for its services. A lease is a particular kind of concession in which the infrastructure costs are borne by the local authority, or group of authorities.

In 1892 in France, a direct management system was in operation in most districts, as is the case today, and the indirect management methods were prevalent only in the most populous urban centres.[22] The larger the district, the more anxious the authorities seemed as a rule to contract out the management of the public water supply, which had been the legal responsibility of mayors since 1884.

In short, the rather muddled systems in operation in most areas meant that the overall situation was complex, confusing even. In any case, the duopolistic position currently occupied by the two major private companies, the Compagnie Générale des Eaux and the Lyonnaise, did not come into being until at least 1920; and in the case of the smaller districts, the lack of financial resources and of management and technical skills in the area of water supply systems were not a major attraction for private companies. At the end of the nineteenth and beginning of the twentieth centuries, municipalities manifestly accepted the problems posed by investment in and management of water supply systems. The reason for this was that, with the introduction of new legislation in 1902, the municipalities received state subsidies, sometimes supplemented by grants from the General Council of the department.[23]

State subsidies in France

The level of subsidy allocated to local authorities from tote funds for the construction of drinking water supply systems were calculated by means of a scale laid down in a decree from the Ministry of Agriculture and communicated to prefects in a circular dated 1 October 1904. These four scales, which were modified in 1921 and again in 1930, were designed for calculating the level of subsidy as a function of the installation cost per inhabitant and in accordance with the degree of affluence in each particular district. In addition, provision was made in 1923 for a special loan in order to make subsidies available to districts where the water supply had been damaged or destroyed in the war.

Thus, by means of a process of social equalization which gave top priority to infrastructure projects of this kind, a supplementary levy on tote funds provided for by the Finance Act of 31 March 1903 (Article 102) and, from 1920 onwards, a second levy on proceeds from gambling were used to finance the installation of water supply systems in poorer districts. However, from 1925 onwards, the rapid increase in requests for subsidies revealed the obvious inadequacy of these special resources, and Parliament made further considerable sums available to the Ministry of Agriculture, either within the framework of the budget laws or of the so-called capital equipment laws, and under the general heading of rural development.[24]

During this period, state subsidies were a gesture of encouragement, and the specific characteristics of each district were taken into account only in order to determine the level of subsidy. They were intended to redress the inequalities caused by wide variations in geographic and socio-economic situations, and also to encourage a new socio-cultural relationship to water, and more particularly to drinking water.

In 1934, after the world economic crisis had struck France and led to budgetary restrictions, the minister of agriculture, Henri Queille, used a very real change in the concept of drinking water supplies to argue in favour of altering the criteria by which state subsidies were allocated. In 1903, it had seemed sufficient, for considerations of hygiene, to supply water only to a certain number of wells and fountains. However, in the 1930s, this level of supply, even though it was far from being realized throughout the whole of France, was considered inadequate; the water supply was now perceived as a public utility that was important not only for reasons of health and hygiene, but also for economic, agricultural and industrial reasons.

The major difference with the earlier period lay in the fact that the 'desirable progress' could only really be achieved when the entire population had a supply of water in their homes. Indeed between about 1900 and 1920, drinking water, which was supplied only to a certain number of public street fountains, was free to the public, since the entire cost was borne by the municipality. Consequently, since financial costs (firstly for capital expenditure and then for maintenance) always resulted in the levying of new charges, it had been standard procedure to examine financial documents in order to extract the most characteristic elements of the situation at local level.

By about 1930, the situation had changed. Local authorities were increasingly tending to establish public supplies of drinking water and to sell the water to consumers; consequently, companies of this kind were not restricted by a rigid framework of purely financial considerations.

The ministerial decree of 29 October 1934 changed the regulations by which subsidies were allocated.[25] From then on, instead of taking into account only the cost per inhabitant supplied with water, which was generally limited to the initial capital expenditure and ignored subsequent operating and maintenance costs, the new regulations were intended 'to reflect reality more closely'[26] by taking into account all the elements of the scheme and assessing it in accordance with a single indicator known as the 'characteristic cost'; this would be an accurate reflection of the total cost per cubic metre of supplying water. Once the 'characteristic cost' had been established, it would be possible to compare various solutions for the same problem and thus to choose the least expensive, for example a gravity rather than a pumped system. This new method of calculation was to remain in force until 1940, modified only by two ministerial circulars, one dated 7 December 1936, which simplified the enquiry procedure prior to the public announcement of the works and the tapping of water,[27] and the other dated 4 March 1937.[28] Finally, on 6 March 1939, the schedules used to calculate the level of subsidy for the installation of municipal water supplies were revised.[29]

From 1900 to 1940, therefore, state intervention in the development of the country's water supply system − as in other aspects of social life and of health − reflected sustained concern with the advantages that could be gained from a good water supply and drainage system. The attractions of safety, hygiene and comfort proclaimed in schools, by doctors and in the media encouraged individual citzens, local communities and the companies to invest in the great consumer product that water had become.

The cost of water

It is extremely difficult, just as it would be for the present period,[30] to conduct an analysis of the price of water in the nineteenth century unless it is carried out at district level or unless an understanding can be gained of the internal functioning of the major companies.

The price of water did have one specific characteristic: instead of arising out of a match between supply and demand, it was fixed by the local authority at a level that would guarantee the financial soundness of the production and distribution service. That being the case, it should be possible to study the process by which the price of water was determined by examining archive files from local authorities (in the case of direct management) and from the companies that worked on their behalf (in the case of indirect management). However, it has proved impossible to gain access to this type of document.

This does not, however, prevent us from assuming that, as at present, the prices charged by private companies were higher than those fixed by community-run enterprises, although the existence of a profit margin only partially explains this disparity.

Secondly, unless it were possible to research local archives, examination of the files on state subsidies, particularly those granted between 1903 and 1934, does not make it possible either to calculate a theoretical price for water by dividing total annual charges by the maximum volume that the network could supply. These files only provide information on projects of which the real cost is unknown, since there is no information on costs once the supply was installed. Consequently, the first of the five items required to calculate the basic price of water is missing, namely the initial expenditure.[31]

It would appear in fact that expenditure varied widely from one district to the next, depending on their situation and the nature of the projects that they were planning. For the 183 districts which submitted proposals to the Higher Committee for Public Health between 1884 and 1890, the average cost was 10.78 francs per head of population.[32] This cost is related in theory to the population served. Its significance is thus not purely financial, but also socio-sanitary. Costs turn out to be very variable when distributed according to size of district: they were high in the case of smaller districts and relatively low for larger towns. As is the case today, costs in small districts, and those in sparsely populated areas, were high. In order to install a minimal system, they had to pay as much, if not more, than average-sized towns which, since they had a high average flow, were also in a position to consider improving the sanitation system shortly afterwards. In large and medium-sized towns, on the other hand, the financial cost per head of population was lower, which enabled these towns to improve or extend their networks, whereas small districts, disadvantaged by the sparseness of the population, had to find ways of financing the heavy initial expenditure required. Finally, the cost per district and per group of districts was higher when the water was taken from springs and rivers.[33].

Calculated on the basis of a sample of districts in twenty-one departments between 1884–1902, the average cost of proposals submitted to the Higher Committee on Hygiene was 24.45 francs per inhabitant supplied.[34] The differences in cost seem to be correlated with the geographical location of the districts and the technical methods adopted. Thus communities in mountainous regions and those in flat areas drained by a dense network of rivers which limited themselves to a basic supply system spent on average much less than the average cost for the sample as a whole. On the other hand, considerable costs were incurred when it became necessary to build aqueducts and reservoirs, to use pumping stations or filtration plants or even use cast-iron instead of earthenware piping.[35]

The profitability of water

Whether high or low, the cost of constructing water supply networks was undoubtedly adapted to the resources available to the various local communities. This was also the case with the management of the networks, which brought a good return for the operators whatever system of management was adopted. By using a certain degree of expertise and modifying at will the price of water to consumers, the 'water sellers', whether public or private, profited from their monopolistic situation. Examination of the annual balance sheets of 217 companies for the year 1894 reveals a positive balance in the vast majority of cases.[36] However, the indirect method of management (concession and company) was generally more expensive for the consumer and the profits made were significantly higher.

In the direct management system, loan repayments were settled by members of the local community (heads of family) through the levying of new taxes. In this way, the initial expenditure required, which was very considerable in large towns (several hundreds of thousands, even millions of francs), was not a burden upon the finances of the water authorities and indeed brought them a fairly handsome return; in 80 per cent of cases, income in 1894 from the sale of water was three times higher than expenditure for the year. For concessions and private companies, income in half of the cases (same sample and same year) was almost six times higher than expenditure;[37] for the other 50 per cent, it was roughly in line with the overall average. Finally, in 1894, the expansion of the market for water was at its peak. Not only had engineers just raised their quality standards and the daily volume of water to be supplied, but the growth in the urban population increased demand and led in turn to an expansion of the 'water business'.

As an expert noted in 1897, revenue increased steadily between about 1880 and 1890 during a period of monetary stability. Thus in the department of Ain, the local company which produced and distributed water in Bellegarde had an income of 1,780 francs in 1878 which by 1894 had risen to 3,340 francs.

Thus in order to meet rising demand, many projects for the extension of water networks were either being planned or implemented, both in major Western cities (in the nineteenth century) and in small towns (between about 1900 and 1930). In other areas, the inadequacies attributed to the infrastructure (dilapidation, leaks, earthenware piping, absence of public or private service) revealed considerable needs, even in towns which already had partial systems.

Thus on the eve of the First World War, the market for water in France was particularly attractive. Technological complications, the relatively slight increase in pollution, except in certain areas (such as the

North) and the wish to get rid of statutory responsibilities, had encouraged only a small minority of local communities to enter into contracts with private water companies; those which had done so included some of the richest and most densely populated areas such as, Paris, Nantes, Nice and Lyons. During this period, and despite the remarkable expansion of a private company like the Compagnie Générale des Eaux, municipal corporations (and then, between 1918 and 1940, groups of local authorities) were the dominant force in the market for water; this method of management was probably adopted in three-quarters of the local authority areas in France. Private companies were thus operating in only 25 per cent of local authority areas; the Compagnie Générale des Eaux was responsible for the water supply in one-third of these areas, with other private companies involved in the remaining districts.

Secondly, the concessions and private companies were not, with a few exceptions, favourably located in large urban centres, but managed water supply networks in areas with roughly the same population as those supplied by the municipal corporations.[38] Initial investment costs were very similar for both methods of management at about 750,000 francs per local authority area.[39] Finally, costs were virtually proportional to the number of people supplied; however, there was a slight advantage in favour of the indirect method, which covered 25 per cent of the population supplied with water but accounted for only 18.2 per cent of capital expenditure, whereas the municipal corporations bore a heavier burden, since they accounted for 81.8 per cent of initial investment costs but supplied only 75 per cent of the people covered. It must be said, of course, that the private companies had to pay dividends to their shareholders.

The prices of water

Although they invested less than their counterparts in the public sector, the private companies nevertheless charged a significantly higher price for a cubic metre of water;[40] in 1894, the price in about twenty districts was 1.03 francs.[41] In the same year, the municipal corporations were charging a much lower basic price of 0.58 francs.[42] This basic price varied from one district to another, sometimes considerably, not only from region to region but even within the same department. The basic price calculated for 1894 ranged from 0.01 francs (Royat, Cauterets, Argèles) to 5.61 francs (Nogent-l'Abesse, Haute-Marne). This gap proved to be much greater than that which exists today in France: in 1894 it ranged from 1 to 561, while the spread today is from 1 to 120.[43]

However, there is no doubt that the water supplied was much cheaper than in the earlier system, and that the biggest reduction was in towns. In Nantes in 1854, the price to the consumer per cubic metre fell from 3.56 francs to 0.82 francs after a new supply system had been installed by the Compagnie Générale des Eaux.[44] In Rennes, it fell from 5.55 francs to less than 1 franc[45] after 1883, again as a result of the construction of a new supply network. The same thing happened in Lyons where, from 1856 onwards, the price charged to private individuals was fixed at 0.60 francs and at 0.30 francs for industrial users.[46] Finally, in Paris, the price per cubic metre fell to 0.40 francs. Despite these reductions, as we have seen, a period of monetary stability meant that water companies could still generate handsome profits, whatever the management system adopted. Indeed, the difference between the basic price and the price to the consumer was quite considerable. It was a margin from which the various taxes and rents were paid, as well as the dividends paid to shareholders. The remainder constituted the profits made by the company or local community. Thus in Nevers, where in about 1900 the basic price was 5 centimes per cubic metre, the price to private consumers was 30 centimes![47] In Strasbourg in 1894, the sale price was considerably greater than the basic price: 20 centimes compared with 13 centimes. In the German Empire in about 1880, there were also wide variations in the price of water to consumers from one city to another: 13 pfennigs per cubic metre in Dresden, 7 pfennigs in Augsburg, 24 in Frankfurt am Main and 31 in Stuttgart, Mannheim and Stettin.

The price of water per cubic metre in the nineteenth century also varied in accordance with the scale by which the consumer paid his dues. There were in fact several methods of paying for the water supplied. In England, there were generally no water meters and water was paid for by a special rate levied on users. Secondly, a lower price could be fixed for certain categories of people and institutions which required large quantities of water. This was the case for industrialists in Lyons who paid half the price per cubic metre charged to private individuals or municipal services (fountains, drains, irrigation, fire service). Thirdly, the price per cubic metre varied according to the pricing regime. Thus at Argèles (Hautes-Pyrénées), water had been sold in perpetuity and was supplied at very low cost to consumers, at least to those who had had enough money to invest in this way. However, during the second half of the nineteenth century, the price charged for a cubic metre of water was not usually related directly to actual consumption because of a lack of gauges and meters. When the annual subscription was paid as a lump sum, there was no charge per tap.

At Nevers, for example, according to a price schedule dated 31 May 1857 – which remained in force until 1909 – daily consumption was estimated by the concessionary company at 1 hectolitre per household, to which were added, as appropriate, 0.40 of a hectolitre per coach or carriage, 0.60 of a hectolitre per horse and 0.50 of a hectolitre for gardens of less than 400 square metres. Baths and breweries were charged at 25 hectolitres, potteries and hotels at 10 hectolitres, inns and laundries at 6 hectolitres and bakeries and cafés at 4 hectolitres.[48] All these estimates of consumption seem to have been purely theoretical. Some customers consumed excessive quantities: one industrialist user charged for 25 hectolitres consumed 80 or 100. It was estimated at the time that 20 per cent of the water supplied was for 'no known purpose'![49] Finally, in Nevers (until 1910), the price was based not on the cost of the water supplied but on the rent paid by the subscriber for his housing; this was also the case in Berlin, Karlsruhe and Frankfurt am Main. In Nevers, this pricing regime, which originated in Britain, meant that for a rent of less than 500 francs, water was charged at 5 francs per hectolitre; for rents between 500 and 1000 francs, the price was 10 francs, with 15 francs being charged for rents above this level. In Paris, the 1881 regulations stated that the annual charge for domestic customers should be based on the number of people occupying an apartment. The disadvantage of this pricing regime, which was a more accurate reflection of reality, was that a large household with a modest income was charged more than a rich person living alone.[50] Thus an office worker with a wife and five children over 7 years of age paid 32 francs for a single tap,[51] while a rich person living alone paid only 16 francs.

This pricing regime did not last long. The use of water meters eventually became compulsory, although not without recriminations. In Nevers, the decision of the 10 and 11 June 1895, taken by the water committee, which fixed the price per cubic metre, was attacked by the opposition on the town council. It was indeed the case that some consumers, particularly large ones, who had been circumventing the regulations, certainly did not wish to see a pricing regime that would involve the use of water meters in order to charge customers by the cubic metre: 'The Association of Chambers of Commerce and Industry protested to the prefect about the Company's outrageous demand and the proposal was not implemented; it was not until 1910 that the use of water meters and a charge per cubic metre were made compulsory.'[52]

This brief examination of the price of water in the nineteenth century enables us finally to draw a few conclusions and to consider several hypotheses. Firstly, the situation in about 1900 was very different from the one we know today. The Compagnie Générale des Eaux and the Lyonnaise, or for that matter the other private companies, did not enjoy the same privileged position that they have acquired today. Nevertheless, the main aim of these companies was to maximize profits. Their persistent refusal to publish their operating accounts can only be explained by their desire to avoid unfavourable comparisons with the

The effects of the conquest

public sector. The few cases studied (nineteen in 1894) showed that the prices that they charged were more than 100 per cent higher than those in the public sector. Even if some of this difference can be explained by the need to make profits and pay dividends, this does not entirely explain the disparity observed between the basic price and the sale price, particularly in the case of the private companies, since the water supply systems operated by local authorities usually also achieved a positive annual balance. Finally, quite apart from the question of the method of management, the essential fact is this: in 1914, most people living in rural France and many of those living in urban areas had not yet experienced the dubious pleasure of having to pay for the water that they used. Except in large and medium-sized towns and cities, in particular prefectures and sub-prefectures, the water supply network was inadequate or nonexistent, and the quality left a great deal to be desired. Consequently, for financial, technical and socio-cultural reasons, drinking water was still distributed only to a small minority, although it is worth noting that it was, as the First World War approached, an influential and active minority.

However, after the body of legislation introduced in France between 1884 and 1914, and more especially between 1920 and 1940, this minority began to increase in number, mainly in towns and cities, but also in the country, as water supply systems in the West changed gradually during the first half of the twentieth century.

8

The development of the infrastructure

The battle for capital investment

As in all the major Western countries, the water supply infrastructure in France was developed first of all in the capital. As early as the reign of Napoleon I and then again from 1860 onwards, the sewer system installed in Paris made the French capital a beacon of civilization, one century after London had achieved the same status. Since they did not enjoy the same degree of prestige, the provincial capitals and other major cities often took a long time to provide themselves with complex and costly systems. Finally, and parsimoniously, rural districts followed suit.

It should be said that the level of investment required was high (see table). Thus it was the major cities (Paris, Lyons, Boston, New York) that first embarked on major schemes between 1760 and 1830. From 1830 onwards, however, the situation changed: although major cities were still committing themselves to major projects, smaller towns, such as Arcachon, Bielefeld and Albany, were beginning to follow their example.

The overall cost of some water supply systems (about 1900, in gold francs)

Paris	24,326,000
Lyons	6,000,000
Bordeaux	4,200,000
Brussels	6,600,000
Philadelphia	3,500,000
Washington DC	1,650,000
New York	67,900,000
Baltimore	36,400,000

The type of works carried out, and thus the borrowing that was

required, was obviously linked to the size of each town: Amiens
borrowed 600,000 francs in 1843 for the 'cost of works relating to the
supply of water in various districts of the city',[1] and Dijon borrowed
220,000 francs 'for additional expenditure incurred in supplying water
to various areas of the city'.[2] In 1844, the town of Chartres borrowed
240,000 francs in order to cover the 'costs of lifting and distributing
water from the river Eure to the highest point in the town'.[3] The
problem of installing the basic infrastructure was then compounded at
the end of the century by that of improving and extending existing
supply networks as the population of town and cities increased. At the
same time, the problem of drainage systems, hitherto a secondary issue,
became a widespread concern. An increasing number of towns and cities
began to pay attention to this question, and even rural communities
began patiently to compile the thick dossiers required by the
Government, providing the Water Engineering Department of the
Ministry of Agriculture with all the documentation required to obtain
the much-awaited subsidies.

These dossiers[4] are a precious source of information for the historian.
They emphasize the link between the chronology of investment in the
water supply infrastructure and the topography, climate, crops and
economy of the regions in question. Some communities clearly perceived
the value of a water supply system, while others were less responsive,
taking into account their economic resources and their particular cultural
habits.

There was a considerable gap between the massive capital projects
required in a major city or a medium-sized town and those called for in
rural communities. In Fenières en Ferté-Chevresis, a hamlet of 983
inhabitants in the department of Aisne, the total cost between 1920 and
1940 was 8,805 francs. In contrast, in towns such as Lyons, Grenoble,
Mulhouse, Saint-Etienne, Dijon, Poitiers and Besançon, the cost in gold
francs and in an earlier period (1830–1850) was several hundreds of
thousands of francs. The sum was considerably higher in the case of very
large-scale projects. The city of Marseilles, for example, arranged loans
on several occasions in order to pay for the construction of the Durance
canal. Ten million francs were borrowed in 1839, 7 million in 1844, 9
million in 1847 and 2 million in 1853. In 1857, the much smaller town
of Metz borrowed 1,446,000 francs for a new water supply system, and
in 1852 Bordeaux borrowed 4,800,000 francs.

As might be supposed, expenditure appeared to vary in accordance
with the size and ambitions of towns and cities. For the same basic
infrastructure (public fountains, water mains, drains), the per capita cost
seems to have been lower in large and medium-sized towns, with a few
exceptions such as Marseilles, which had to finance the construction of
the Durance canal and which occupied an unfavourable geographical

location. It should also be stressed that this expenditure was financed by loans secured against new taxes and that local authorities, despite their desire to invest, were very often reluctant to appeal to their citizen electors. Moreover, it was clear that what might be called the 'hierarchy of cities' was maintained, since the major cities were the first to invest in drainage and water supply systems.

In 1850, Besançon borrowed 700,000 francs at 5 per cent, repayable within sixteen years from 1853 out of its normal revenue. The loan was used to cover part of the cost, a total of 1,250,000 francs, of 'drawing off and supplying water from the Arcier spring to the various districts of the town and of completing the construction of the drainage system already in progress'.[5] Since the town had no existing debts and its annual financial surplus was 131,308 francs, its application was immediately accepted and its indebtedness readily countenanced.

On the other hand, the city of Beaune, which was already heavily in debt and which was forced in 1849 to increase taxes in order to cover its deficit, was able only to install public fountains, while at the same time increasing its indebtedness; the city borrowed 150,000 francs at 5 per cent over ten years, which it began to pay back from 1852 out of increased taxation.[6] For the same reasons, Vannes planned in 1856 to

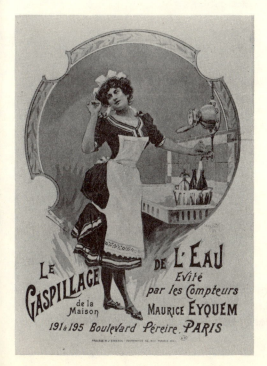

1898. Water starts to flow from the taps . . . in the larger towns and cities. By discouraging wastefulness, water meters fitted easily into the spirit of parsimony that permeated bourgeois morality.
(© Bibliothèque des Arts Décoratifs, Paris. Photograph: Éditions Robert Laffont.)

construct only one water main between Meucon and the fairground.[7] At the same time, on the other hand, Grenoble was already in the process of borrowing in order to build 'various drains and sewers'.[8]

The early start made by larger, more prosperous cities gave them a considerable lead over other towns and cities and, particularly, over the whole of rural France. At the end of the nineteenth century (around 1880–1900), they were improving their water supply networks and finishing the construction of their sewerage systems. This was true of Caen, Calais, Castres and Béziers in 1888[9] and was made necessary by the rapid increase in the urban populations and the number of medium-sized towns with between 20,000 and 50,000 inhabitants.

However, the dossiers that provided the information on which the ideas outlined above were based related only to local authorities that received loans from the state. In fact, inspired by the frequent and well-publicized pronouncements of hygienists and the widespread fear of cholera that prevailed between 1830 and 1850, a small number of local communities had taken a lead earlier in the nineteenth century and begun to provide public drinking water supplies.[10]

A survey of 691 towns and cities with more than 5,000 inhabitants carried out in 1892 by the engineer Bechmann reached a similar conclusion[11] by noting the periods at which the existing water supply systems had been constructed (see table).

Towns and cities with a water supply

Period	Number
Before 1700	7
1700–1800	8
1800–1820	4
1820–1840	23
1840–1850	11
1850–1860	55
1860–1870	94
1870–1880	74
1880–1892	92

Calculated on the basis of the annual average per decade, the rate of increase was spectacular. The first thing to note is the immense difference between the eighteenth and nineteenth centuries. Then during the nineteenth century, there was a marked acceleration between the first and second half of the period, despite a temporary slowing down between 1870 and 1880, due probably to the fall of the Empire, the Franco-Prussian war and the Commune.

However, it should not be concluded from this that the home of every French citizen was connected to a water supply and a mains drainage system. Firstly, until about 1900, virtually no homes had their own private water supply, except for those of a small minority of privileged people. Secondly, only a very small proportion of the population was connected to the network. . . even when one actually existed! In fact in 1892, only 290 French towns out of 691 were supplying water under pressure and distributing it to their consumers. These 290 towns had a total of 4,512,941 inhabitants but only 127,318 subscribers. Even if it is assumed that public institutions (boarding schools, *lycées*, barracks, convents, hospitals) were counted as a single subscriber, the proportion of homes supplied was still very low.

Bechmann's survey also shows that drain and sewer systems were much less well developed than water supply networks. Only 90 towns out of a total of 691 had sewers, and only 156,054 subscribers were connected to them. Finally, in the opinion of an engineer like Bechmann, waste disposal left a lot to be desired. In the vast majority of towns, sewers were discharged directly into rivers and streams: this was the case in 354 towns of a total of 420 which replied to the survey on this point. Only 17 towns (total population 225,913) out of 313 (total population 7,158,316) had a system of drainage into the sewers, while the majority (294 towns with a total population of 6,918,229) used cesspits and a tiny minority had mobile latrines (2 towns, with a total population of 14,174). For a very long time to come, both the water supply and waste disposal systems were to remain problematic and of poor quality, even nonexistent.

There were several reasons for this state of affairs, which was considered deplorable by hygienists of the period. Firstly, 'in France, hygiene is not a profession.'[12] Medical students could not see the value of the notions squandered on them in hygiene lectures: comparative pathology, animal experiments, chemistry, industrial technology, epidemiology. Ignored in favour of clinical medicine, pathology and physiology, hygiene seemed to them to be merely a pretext for philosophical, humanitarian, even political declarations which had little to do with scientific medicine. Because the study of hygiene in France did not lead to any vocational opportunities, and also because it was not profitable, either in general medicine or social medicine – which did not begin to develop until the 1890s – it occupied a marginal position. This was not the case in Great Britain or the USA.

Secondly, 'hygiene committees do indeed exist in certain towns in France, but they meet only once a month at most, if they meet at all.'[13] In 1879, departmental hygiene committees often existed only on paper, as was the case in Canada in a slightly earlier period. Only a few functioned properly. Their reports constantly highlighted the refusal of local councils to grant them a budget to publish their studies and to pay their members' attendance allowances.[14] Several of these committees did not even have a secretary: all the documentation was dealt with by a clerk in the prefect's office, who received and passed them on without allowing committee members time to study them.

The legislation then in force, which was liberal and concerned in its intentions, was hardly implemented at all in areas of public health. This had a considerable effect on the provision of supplies of clean water and efficient sanitation systems.

In fact, for almost the whole of the nineteenth century, public health was only of theoretical interest to the government. It did not really come into its own until a more suitable level of development had been reached and the state began to take increasing responsibility for public health. This was not the case in other Western countries, at least in urban areas, where the level of provision was significantly more advanced at any given period. Thus in 1911, 96 per cent of dwellings in London were connected to a water supply, compared with a figure of 17.5 per cent for Paris. In the major European cities (including London, Paris and Berlin), consumption per head of population per day was only 86 litres, compared with 341 litres (155 gallons) for major cities in the USA (including New York, Philadelphia, Baltimore, Chicago and Detroit).

Another significant difference was that by 1895 about 42 per cent of urban communities in Prussia with a population more than 2,000 had a water supply network; in 1930, only 25 per cent of communities in France with more than 3,000 inhabitants had such a network. As for the United States and Germany, they had a significant lead over France by about 1900 in respect of the percentage of the population served by water supply and sewer systems. Moreover, progress in those countries remained more rapid until about 1930. Finally, there was still a wide gap between the advantaged urban population and the disadvantaged rural population, and between developed and less developed regions (for example, Northern and Southern Italy).

The case of towns and cities

The Parisian 'model'

In France, despite the interest expressed by the medical profession, the views of scientists often remained a dead letter. The most obvious exception was the city of Paris. In about 1800, Paris had some 600,000 inhabitants and was supplying 8,000 cubic metres of water per day, or

1.33 litres per head of population; the city had 25 kilometres of sewers, or 1 kilometre for every 23,000 inhabitants. In 1900, the year of the Universal Exhibition, the population of the city was about 2,700,000 and water consumption was 249 litres of water per head of population per day (wastage was estimated to be about 50 litres). There were 1,113 kilometres of sewers, about 1 kilometre for every 2,240 inhabitants.

This was why the authors of the *Annuaire sanitaire de France* (French Sanitation Yearbook) exclaimed so enthusiastically that 'One of the greatest achievements of the nineteenth century will be the creation of this great sanitation system, the functioning of which does great honour to those who designed and constructed it.'[15] It was indeed during the second half of the nineteenth century (after 1854) that the decisive progress was made under the administration of the prefect Hausmann and at the instigation of the great innovator and organizer Belgrand.

Prior to this period, the system had been extended slowly and without any changes being made to the principles and mistaken ideas of the past. Girard had increased the volume of the water supply through the construction of the Canal de l'Ourcq, and the reconstruction of the Chaillot fire pump by Dupuy had had a similar effect. For his part, Mary had improved the supply by building a certain number of reservoirs. Emmery, Coïc and Dulan had developed the sewer network, and Dupuy had introduced the ovoid type that had recently emerged in England.

However, it was in about 1860, during the administration of Belgrand and his successors, that two major innovatory principles were put forward and implemented. The first was to improve the quality of the water supplied, which was at the time rather poor, and the second was to push sewer outlets out into the river and, to this end, to reroute the sewers away from the natural slope of the land.

For domestic use, water from the Seine and the Canal de l'Ourcq was replaced by spring water, the limpidity and freshness of which was much appreciated. The sewer outlets were then taken outside the city boundaries to a position well downstream in the Seine valley in order to await new measures, such as the halting of discharges in dry weather and the introduction of large-scale treatment of sewage by spreading it on agricultural land.

Two principles were put into practice by Belgrand. Firstly, he advocated a dual water supply system which would completely separate the supply to dwellings (which was kept as pure as possible) from the supply to streets and factories, which could be of lower quality. He planned for two completely separate supply systems in the capital, one serving domestic users and the other supplying industrial users and public places (public highways, avenues, courtyards and gardens, stables and coach houses, factories, etc.).

Secondly, Belgrand outlined in a masterly fashion a system of sewers intended to collect all the waste water from the various districts of Paris and to take them by the shortest route to a carefully selected and sufficiently distant point on the river bank, where it would be possible to treat it. Mille, who was given the task of researching this last part of the scheme, found a solution in agricultural irrigation, an idea which was enthusiastically supported by Alfred Durand-Claye.

By about 1900, therefore, Paris, unlike many European cities, had opted, rather later than most, not for the purifying system but rather for a combined sewer system. This choice, made by Belgrand, was hotly disputed by several of his colleagues. And yet, the separate system which, on the one hand, keeps household water and faeces apart and, on the other, isolates street water from rain water, does have considerable advantages. Firstly, it is economical to install, since the relatively low and constant rate of flow means that the draining of domestic waste water and sewage requires only narrow pipes. Secondly, the water in the general main is usually not too dirty, which means that it can be discharged directly into the river or purified at low cost.[16]

In Paris, on the other hand, everything – street water, rain water, solid waste from the roadway, waste water and sewage – went into the sewer, at least from those dwellings that were connected to the mains drainage system at this period. The characteristic that distinguishes the Parisian system from that in other European capitals is that it is made up of a series of subterranean pipes of considerable diameter, since they were intended to accommodate water mains, telephone cables and other utilities. (Moreover, this drainage system posed the problem of how to treat the waste water.[17] What was to be done with solid waste? Should it be disposed of via a main sewer, or physically or chemically purified?)

The solutions adopted by Belgrand were not exempt from criticism at the time. The main complaint was that Paris had opted for a dual water supply network, one for domestic and the other for industrial users. Contamination of domestic water by the less pure industrial supply did indeed cause some intestinal problems for Parisians at the end of the nineteenth century.

Nevertheless, in 1900, the achievements of Belgrand and his successors aroused both admiration and envy from the many administrators, politicians and hygienists who saw the way in which Paris had solved the problems of water supply and waste disposal as a model for Western civilization. If in 1850 the French capital had lagged far behind other major European capitals in this respect, by the time Belgrand died in 1878 and, more particularly, by 1900, considerable progress had been made.[18]

From 1882 onwards, the extension of the sewer network was accompanied by the development of the mains drainage system of sewage disposal. This system, which was already in use in several foreign cities, including London and Brussels, was then unknown in Paris. The severity of the typhoid epidemic in the second half of 1882 led the city authorities to concern themselves with the problem of sewage disposal. By a decree dated 25 October 1882, the prefect of the Seine set up the Technical Committee for the Sanitation of Paris which was entrusted with the task of finding 'the method of disposing of faeces most in accordance with the laws of hygiene' and of indicating what changes should be made to the sewers, to the handling of waste water and the disposal of refuse. A study trip was arranged to those capitals considered to be the most 'hygienic', Brussels, London and Amsterdam. Once it had been ascertained that mortality from typhoid fell significantly when a mains drainage system was installed, the Committee, after many studies and despite violent criticism, accepted that sewage could be purified by the soil.

The works began in 1883: progress was slow until 1886, but accelerated from 1890 to 1910, considering the rate of increase in the number of direct discharges into the sewer: fewer than 500 for Paris in 1886, 4,667 in 1890, 23,055 in 1900 and 49,010 in 1910.

Thus a mains drainage system and a drinking water supply were eventually installed in Paris, though not without difficulty.[19] Landlords were violently opposed to the changes, since they feared additional costs and taxes, and also contagion; however, it was evident that these new installations were simply increasing the rent that they could charge for their properties.

Large and medium-sized towns and cities

The development of the water infrastructure in large and medium-sized towns and cities clearly shows that Paris was not only a model but also an exception. These towns and cities were all slow to make the necessary investment, and the works that were carried out, often late and in piecemeal fashion, were hardly well suited to the growth in population nor to the general economic expansion.

Marseilles   In this respect, Marseilles[20] was an exception. Its greatest hygiene problem was that of water, and as early as 1821 the city council drew up plans for a canal scheme. On 12 July 1834, the council, at the instigation of mayor Consolat, stated that 'the construction of the canal is an irrevocable decision; whatever may happen and whatever the cost, the canal shall be built'; and *Le Sémaphore de Marseille* supported the scheme in its edition of 15 July 1836. The canal, a massive and lengthy project, was the work of the engineer Franz Mayor de Montricher. Eighty-two kilometres in length, sixty-seven of which were open to the sky and fifteen in tunnels, the canal required a massive labour force, both French and foreign. Work began in 1838, and the inauguration ceremony was held on 29 November 1849 before 15,000 to 20,000 spectators. However, the sanitation system was not constructed until the end of the nineteenth century. Because it had not been planned to serve the over-populated and sprawling suburbs, it soon proved to be inadequate.

Angers   In Angers,[21] the insistence of several doctors and meetings of the local medical society were not sufficient, even just after the cholera epidemic of 1832, to persuade the conservative and parsimonious local authorities to carry out the major works required for the installation of a water supply and sanitation system. For twenty years no action was taken. There followed interminable discussions of the respective merits of water from the Maine and that from the Loire, a point on which a clear decision could not be made on the basis of the analyses conducted.

Work did not begin until 1854, the year that cholera returned. The supply, which was at first restricted to a hundred or so street fountains, was gradually extended to the homes of subscribers after the installation of a second steam engine. From 60 litres per day in 1856, the estimated consumption per inhabitant increased to 150 in 1860 and to 270 in 1892; it then fell to 100 litres in 1900 before rising again to 182 in 1933. These figures show that the consumption level per head per day reached in Paris in 1900 was attained in an episodic manner and only after a long delay. Once again demand, particularly from horticulturists and market gardeners, rapidly exceeded supply. In 1907, about half of the households in town, 5,700 in number, were subscribers; this represented a very good rate of progress. The supply was installed from the centre towards the periphery and in the nineteenth century remained strictly limited to the local authority district.[22] It took a long time for the city to equip itself with a sanitation system, which was equally costly. Thus the old practice of discharging everything into the river persisted for a long time. Installed with great reluctance and delay and with inadequate funding and intended primarily for the well-to-do districts in the city centre, the sanitation system in Angers was typical of a conservative option that gave low priority to public health.

Bordeaux   In some respects, sanitary policy in Bordeaux is strangely reminiscent of that in Angers. In Bordeaux, however, it was the major works carried out by the *intendants*, or provincial administrators, and continued during the First Empire which determined the precise social localization, particularly in the Chartrons district and in part of the parish of Saint-Pierre,[23] even though the city expanded considerably in the nineteenth century. Just as in Paris and in many other French towns and cities, the First and Second Empires were periods in which major urban projects were accomplished. The hazards that threatened Bordeaux because of its location and soil had been condemned as early as 1810; this led the local authorities to express their desire to eliminate 'pernicious fevers' from the city and to combat the 'foul-smelling gases' that emanated in summer from the marshes and rubbish heaps.

During the First Empire, the streams were cleaned out, the ditches of the old fortifications were filled in and the first sewers built, notably in the middle-class district of Chartrons. Much concerned with economic liberalism, municipalities during the period of the parliamentary monarchy used the fear caused by the cleaning out of the Pengue as a pretext for not finding the resources required to improve their city's sanitation, despite the wishes of the opposition and the local medical society. Consequently, with the return of the Empire in 1852, old plans were brought out of boxes in order that favourable economic and political circumstances could be exploited; the municipality got heavily into debt, not without encountering strong opposition in its midst. It was during this period that the city authorities decided to take water from springs and outlined a sewer system. However, the water supply problem was not resolved until 1887 after seven years of work and the expenditure of a considerable sum of money (5 million francs). During the same period, a sanitation system was also installed. Bordeaux is located on a site that does not lend itself to the development of a large city, and the sanitation system was not installed until very late. There was a considerable gap between the invention, and the subsequent innovation and its diffusion, a gap that was further widened by the reluctance of social conservatism and economic liberalism to see the old order changed and its power and influence diminished.

Rennes By the end of the nineteenth century, when the city of Rennes had expanded across the river Vilaine, its site was hardly any more favourable than that of Bordeaux.[24] However, the installation of a proper drainage system had been largely completed during the first half of the century and the Second Empire. During the First Empire, the sanitation of the Mail district and the outskirts of Bourg-l'Evêque had been improved. During the reign of Louis-Philippe, the river Vilaine had been canalized, while at a later date the building of the district around the station had led to an arm of the river being filled in. The problem of a drinking water supply was solved at the beginning of the seventeenth century, but surfaced again between 1720 and 1880.

The architect Gabriel remodelled the layout of the water supply system. However, the work, which was started in 1727, came rapidly to a halt and was not resumed until sixty years later.

A fire at the beginning of the eighteenth century had destroyed the earlier supply network. Although Gabriel began to rebuild the system in

1727, the water supply network did not become a reality until the end of the nineteenth century.

In 1848, Marteville wrote in his *History of Rennes*:

In its present state, the city has no water supply system of any kind, either of drinking water or for irrigation, and we cannot see how or when it will escape from this incredible situation!

It was only after 1860 that the problem was taken up again. Among the various projects that were considered, the architect Martenot's proposal that water should be taken from Minette and Loyzance springs was the one selected. On 25 April 1874, the municipal council adopted the plan and, on 13 April 1878, it voted through a loan of 4 million francs for the works. On 14 July 1882, Le Bastard, the mayor of Rennes, inaugurated the first water supply system. As in Angers thirty years earlier, the taste for spectacular success combined with the need to show taxpayers that their local taxes were being properly used led the municipality to construct a temporary fountain in the square in front of the town hall, which was subsequently replaced by a more lasting structure in the Place du Palais.

The drinking water supply was not completed until 1882. The system used the 32 metres difference in height between the source and the highest point of the city to force the water by gravity along 45 kilometres of pipes. It was stored in a reservoir with a capacity of 15,000 cubic metres, which was supplemented in 1889 by a second reservoir with a capacity of 10,000 cubic metres and in 1919 by a third with a capacity of 27,000 cubic metres. The average daily volume supplied in 1883 was 12,000 cubic metres, or about 180 litres per inhabitant per day. This first supply system met the needs of the population of Rennes until 1933. Nevertheless, it soon became inadequate: in 1931, because of the increase in the population, the daily volume supplied was only about 135 litres per head. In 1933, a new treatment plant was brought into operation at Mézières-sur-Couesnon which treated a further 15,000 cubic metres per day. As far as the supply network was concerned, it increased in length from 37 kilometres in 1882 to 82.6 in 1919 and almost 100 by 1939. This made it possible to supply an increasing proportion of the local population. Thus by 1939, the city had 25,000 cubic metres of water available to it each day, or about 400 litres per inhabitant per day. In fact, leakage from the pipes gave rise to considerable losses, to say nothing of the seasonal fluctuations which reduced water pressure and led, in 1939, to the supply to the upper floors of houses in the town centre being cut.

Once the water had been brought to the city and distributed, it then had to be drained away and, if possible, purified. A report by the civil engineer Brière, submitted in 1875, revealed the harmful effects of the absence of any general sanitation system: the city's sewers, which were too numerous and badly planned, were poisoning the well water which the inhabitants used for domestic purposes, including cooking, since the waste water and sewage carried by these dilapidated sewers were discharged straight into the Vilaine without having been purified. Logically, at the request of Dr Perret (1874), the first sanitation system was declared a public utility in 1887, shortly after the construction of the first water supply network. As in Angers, Paris and Bordeaux (at different dates), it served only the town centre. Again like Angers, the drainage system was built only very slowly: by 1890, 27 kilometres had been built, in 1919, 51 and in 1933, 69. Linked to two main sewers on either side of the Vilaine, as in Paris, the drains then led to a general sewer which discharged waste water into the Vilaine two kilometres away from the city. Again like in Paris, the system that was installed in Rennes was a combined system which dealt with both rainwater and waste water together. Since it conveyed these two sorts of water to the treatment plant, the system had the disadvantage of considerably increasing the volume of water that had to be purified. The Parisian model was thus very significant, and the conquest of water took place only very slowly, as was also the case in Nevers.[25]

Nevers In 1827, the mayor of Nevers undertook to provide his city with a water supply system; his plan was to supply 600 cubic metres of water from the Loire at a rate of 10 litres per inhabitant and 30 litres per horse and he outlined the considerable advantages of the scheme 'with respect to the salubrity of the city, particularly during the heat of the summer, to comfort, to the cleanliness of the streets [. . .] and to the assistance provided by fountains in case of fire'.[26] The scheme was not of course entirely untainted by electoral considerations and, once duly elected, the mayor was obliged to implement it. The works were completed in three years. The system comprised a 12 hp steam engine capable of delivering 600 cubic metres of water to a reservoir with a capacity of 650 cubic metres situated 32 metres above the Loire. However, the distribution system did not function perfectly. The engine was low powered and unreliable, and the supply was by no means comprehensive: only almshouses, the barracks and a few factories were connected to it. Entire districts had no supply at all, as was the case in Rennes, Angers and Bordeaux. Thus in 1842 it was decided to extend the network and install a further twenty street fountains. However, even after the expansion of the system, only a very small proportion of the population was connected to a network which by 1857 was made up of only twenty-four fountains and a total length of 4 kilometres. For thirty years, this was the price that had to be paid for early investment. The people of Nevers had to be content with river water of deplorable quality, muddy, silty and full of substances in suspension, too warm in summer and foul smelling. The openings of the purification tank became blocked every time the Loire was in spate, and it was clogged with silt and mud that eventually cut it off completely from the drain tank. Consequently, the system had to be shut down in order for the tank to be cleaned out. Finally, the water intake was located below the place where washing was normally done, 300 metres downstream of a floating wash-house, and at the outlet of the Nièvre, a veritable sewer serving the whole of the lower part of the city. Purification procedures were thus rudimentary!

Despite all these difficulties, the consumption of water by public services and the few private individuals connected to the system tended to increase between 1830 and 1860. As a result, plans were made to use spring water, which was said to be purer. However, mistakes were made. It was not yet possible to calculate precisely the yield of springs, and the estimates made by an engineer from the Highways Department were completely wrong. Not only were aqueducts 3.8 and 7.9 kilometres in length built to convey the water, but also, once the works had been finished, it was realized that these springs could supply only half the water required. As a result water once again had to be taken from the Loire and a new steam engine installed that could lift 800 cubic metres per day. In view of the increase in consumption, it was planned (as at Rennes) to construct a new system capable of meeting the demand projected for 1886. The plan was not implemented until 1910.

For half a century, water consumption remained static, despite the increased population, while the quality of the water still failed to meet even the standards of the period. Firstly, the spring water brought to the city after 1860 proved to be less pure than anticipated, since it was polluted by manure from stables situated near the water intake; in rainy periods, the water was very suspect. No precautionary measures were taken in this respect until 1908. Secondly, the new water intake, which had been installed in 1860, was located downstream of two gelatine factories, floating wash-houses and the sewers of Nevers, which made the mouth of the Nièvre an enormous sewer several kilometres in length, 'a stinking mire, a foul swamp, the mud of which is in view virtually all year round, revealing the filthiest rubbish scattered around: kitchen refuse, the corpses of animals, waste of all sorts, to say nothing of the lavatories that empty into it'.[27]

Thus in 1910, barely 3,000 of the total population of 27,000 were supplied with suspect water from the Loire which underwent no bacteriological examination whatsoever. The rest of the population drew their water from about 350 wells which were often contaminated by organic matter and located near to leaky septic tanks. In 1912, the Higher Committee for Public Health in France declared the water supply to be 'inadequate in quantity, quality and extent, [. . .], the sewers inadequate', and the network 'very incomplete',[28] particularly since waste water was discharged untreated into the Nièvre and the Loire.

In 1909, aware of these deficiencies, the city authorities drew up a plan which was not be fully implemented until 1923. Eight filtration wells made of reinforced concrete were built upstream of the Nièvre (but downstream of the headrace and a glue factory). A system of bacteriological examination was set up, while private wells slowly fell into disuse (they did not disappear entirely, however, until 1950). In Nevers, as elsewhere, the problem of water determined the rest of sanitation policy. And by about 1900, the city did not have a satisfactory sewer network, a purification system for waste water or leakproof cesspits. In Nevers itself, it was not until 1913 that it was decided to to do away with the septic tanks that had been condemned by Calmette. And even in 1946, only 80 per cent of buildings in Nevers had a supply of water, and less than half were connected to the main sewer.

Limoges Unlike Nevers, Limoges did not experience the problems attendant upon the premature construction of a water supply. In 1860, a journalist described the sanitation situation in the city in the following succint, cutting manner: 'Limoges is a sewer.'[29] Because of a lack of water, the city of Limoges was unable to drain its subsoil, clean its streets (at least not until 1877) or to take effective action against the fire of 15 August 1864. The first large-scale urban projects were not undertaken until 1870, under the direction of mayor Othon Pécounet. The water supply was improved and a network of subterranean sewers constructed into which the individual outfalls drained. The use of cesspits was abandoned.

Toulouse On the other hand, Toulouse, like Rennes, had been concerned from the Middle Ages onwards to have a good water supply and was one of the first large French cities to benefit from a modern water supply and distribution system. This was made possible under the provisions of the will of Lagane,[30] its former *capitoul*, or municipal magistrate, who died in 1789. When his widow died in 1817, a legacy of 50,000 francs became due. The municipal council immediately decided to use it to install the first public fountain. An open competition was held, which was won by the scheme proposed by the Toulouse engineer Abbadie, which provided for a supply of 80 litres per person per day at a total cost of 80,000 francs. Thanks to the legacy of its former *capitoul*, the city authorities had to approve a loan of only 30,000 francs. They used the legacy to provide the city with the latest technical advances: a so-called twin-geared engine was installed: if one of the two gear trains needed repair, the other one came into operation to sustain the water supply.

This hydraulic machine[31] was a combination of high-quality lift and force pumps; it was located in the water tower on the highest point in the city in order to provide consumers with water at an adequate pressure. Between 1821 and 1829, a series of three filters was installed in the alluvial banks adjoining the embankment where the water tower was situated. The water that was taken from the Garonne to the drain tanks in the water tower was thus filtered before it reached the tanks. A network of subterranean pipes conveyed the treated water to three monumental fountains. From there, the water was taken to the various parts of the city by a network of pipes. In this way, the water was conveyed to 96 fountains situated in lower parts of the city.

Toulouse was ahead of most other cities of its size in not restricting itself to a public distribution network. It immediately planned the construction of a supply network to private households and in 1827 obtained the necessary authorization by royal decree. Then in 1829 the city set up a technical department consisting of an engineer and two fountain-makers.

The limited number of concessionaries of the 1830s and 1840s gave the impression of being a privileged class. Once again, however, this enviable situation, which in theory made available to the citizens of Toulouse 80 litres of water per inhabitant per day from 1829 onwards, did not last.[32] Despite a rapid increase in population, resources were not expanded accordingly and the quantity available fell by half between 1832 and 1859. Faced with this situation, which was considered dangerous to public health, the city council in 1864 approved the loans required to purchase and install a new hydraulic system in the water tower which could lift 1,000 inches of water in twenty-four hours. At the same time, 'in order to provide work for the working classes', urgent work was begun on the digging of new filtration galleries;[33] this was completed in Februay 1869. Secondly, the municipal council decided to extend the water supply network to the right bank of the Garonne, 'in view of the considerable increase in the city's population'. To this end, new transportable engines were purchased in August 1869 and installed in the Bourrassol station. It was also decided to start work on an 8,000 cubic metre reservoir for the storage of 'fresh, pure water'. However, the new galleries that had been constructed were not able to supply the quantity required; instead of the 1,000 'inches'[34] planned for, the filters had an output of only 350 inches in 1865 and 700 by 1870. Moreover, the terrible flooding of 1875 caused extensive damage to the new filtration galleries which, even after repairs had been carried out, still provided cloudy water with a poor taste. New schemes were then decided upon. There then ensued a veritable race between the growth of the city and the expansion of the water supply system; this lasted from 1871 to 1902 and started up again from 1929 onwards. The system of filtration galleries, which was inadequate in summer, was replaced by a system by which water was taken directly from the river and filtered artificially.

The wish expressed by the former *capitoul* in his will of 1788 was thus fulfilled in three phases, about 1830, 1880 and 1930. Toulouse was thus one of the first provincial capitals to have that 'pure, pleasant tasting water' that he wished for his fellow citizens and of which he regretted the disappearance at the end of the Roman era. In 1838, Stendhal celebrated this water in the following words: 'The water in Toulouse not only has the supreme beauty of the water that one drinks in Rome, it also has the same light and agreeable smell.'

The investment in water supply systems was thus carried out in phases. The First Empire and the Restoration saw plans drawn up and work begun, while the Second Empire saw the first projects under way in Paris and several other large cities. During the *belle époque*, some cities which had been lagging behind (Limoges, for example) and cities with developed systems (Toulouse) started or completed their investment programmes. Finally, in the interwar period, considerable technical progress was made and earlier schemes were extended, renovated or replaced.

The most important observation to be made, both with respect to the chronological evolution and the installation of water engineering equipment, relates to the links that existed between water and power.

The correlation between periods of strong government (both politically and economically) and the expansionary phases in the development of water supply systems is symptomatic. There is nothing surprising about this. The power of a country depends to a large extent on the satisfaction of 'needs' that the conquest of water helped to ensure in the areas of public health, agricultural and industrial production and trade.

This is no doubt the reason why public institutions such as hospitals, prisons and schools were the first to obtain a water supply. It also explains why the modernization of public hygiene was concentrated for a long time in urban areas, and more particularly in the well-to-do districts where the dominant class resided. And finally, it accounts for the monumental character, whether visible or not, of the projects that were realized, mainly after 1860. This can be seen not only in ostentatious aqueducts and splendid fountains (in Paris, Montpellier, Toulouse and Angers) but also in the design of water towers, reservoirs and treatment plants, which were always inaugurated with great ceremony and symbolized the industrial producton of water in the nineteenth century.

With their smattering of Greek and Latin culture, mayors, members of parliament, doctors, architects and engineers, whether in France, Western Europe[36] or North America,[37] were often inspired by the Roman model.[35] When Bechmann, at the beginning of the nineteenth century, studied the supply of water in antiquity, he expressed great admiration for Roman water engineering.[38] And in 1843 Teissier-Roland wrote: 'A well-constructed aqueduct with solid foundations will not require any major masonry repairs for 2,000 years; this is proven by the example of all the Roman aqueducts.[39] They demonstrated their technical competence by undertaking completely new projects and calling into question the old water supply systems. From the social point of view, their aim was to provide each household and each individual with pure water, while at the same time providing a drainage system for waste water. Finally, from the economic point of view, the local authorities and then, after 1902, the state, devoted considerable sums of money to this end.

Although it was a slow and costly process, the widespread provision of running water had two advantages. Firstly, it provided a growing population with good-quality water. Secondly, it did so at a price that was generally acceptable during a period of economic expansion. Consequently, it could be genuinely considered as a social conquest.

The country

Very few rural areas in France possessed modern water supply systems before the end of the nineteenth century. Even fewer could boast a drainage network for waste water. For a long time, waste water and sewage were just discharged straight into the street and into rivers.[40]

Nevertheless, there was no lack of concern with obtaining a supply of pure water. The water from certain fountains and wells was popular among villagers by virtue of its pleasant taste, while the water from others was considered excellent for cooking vegetables or for washing clothes. Even water to be used for hygiene and cleanliness was not wholly absent from their concerns, though little was used for these purposes.

The supply and drainage systems were several centuries old and often maintained with care. Although modest, they were vital to the life of the community, particularly in stock farming areas where the water supply was a constant preoccupation. In the department of Doubs, the need to supply 30 to 40 litres per head of cattle per day encouraged many rural districts from the 1860s onwards to install a carefully maintained water supply. Domestic and agricultural needs thus combined to persuade peasant farmers to accept without grumbling the construction of a wash-house or a combined fountain and drinking trough. Sometimes they not only made contributions out of their own pockets, but also lent a team of oxen or the strength of their own arms to assist in the repair or reconstruction of such structures.

Nevertheless, these installations had changed little by the *belle époque*. All villages had a water point, even if it was some distance from the houses. Most of them had a wash-house and a few had a separate drinking-trough.[41] However, these villages were usually fairly large, almost small market towns.

Thus the water supply network was a typically urban phenomenon. The initial cost and subsequent maintenance costs put it beyond the resources of rural communities. Wells and fountains, even though slightly modified, remained the main source of water in the countryside. Wells were fitted with a solid lip, while the bucket was no longer always made of wood but often of zinc and the rope was replaced by a metal chain. At the request of some enlightened person, the water was sometimes even analysed by a local pharmacist. Once the notion of hygiene had proved its effectiveness, it won the support of prudent farmers concerned to safeguard the health of their animals. Thus source works were often improved, with the wooden pipes being replaced by a more solid material and an iron grille put up around the edges of fountains in order to keep out animals.

On the other hand, drainage and sewer systems were unknown. The removal, transport and disposal of faeces remained the prerogative of several larger market towns which had adopted city habits.[42] Elsewhere, excrement was 'recycled'[43] for use as a fertilizer on fields and vegetable gardens. If it was not used in this way, it was absorbed by the soil and subsoil. Pools, rivers, streams, underground water courses and well water were natural recipients for it. Leakproof tanks and cesspits designed for the storage of human and animal waste and excrement were few and far between. To the alarm of hygienists, wells were often located adjacent to dunghills. To peasant farmers, there was simply nothing horrible or morbid about excrement. The notion of pathogenic microbes remained alien to them for a long time; particularly since in their experience excrement had always had a beneficial, even life-giving effect on the growth of vegetables.

In spite of everything, peasant farmers gradually changed their long-established habits during the first half of the present century. Many mayors sought and received advice from hygienists at department and district level. Many people in rural areas were aware of the shortcomings pointed out from 1900 onwards by the Council for Public Health in France: the inadequacy of the water supply, both in terms of quantity

At Gressigny (Puy-de-Dôme), the fountain was the main focus of village activity. It served both men and beasts and revealed the sexual division of labour. Men led the animals to water, while women did the washing.
(Photograph: © Roger-Viollet.)

and of quality, the remoteness of water points and the absence or poor condition of installations. Thus after 1902, an increasing number of mayors, encouraged by the promise of government subsidies, compiled with touching dedication the thick files that would enable them to supply their communities with water of good quality. These files contained all the necessary information, from chemical and bacteriological analyses, projected average yield and the type of head race proposed, to the finance required.

Contrary to an opinion that is too widely held, this modernization was not restricted exclusively to towns and cities. According to a sample survey for the year 1892,[44] the majority of rural communities had fewer than 1,000 inhabitants. However, it was precisely these communities that requested that their water supply systems be modernized.

In the department of Ain, market towns and nearby hamlets installed new systems between about 1910 and 1930. Thus Versonnex, a hamlet of some 161 inhabitants, was no longer content to use 'pumps located near stables, cowsheds and dungheaps [. . .] which constituted a real risk to public health'.[45] The community planned to take water from two springs at a rate of 2 litres per second, which would provide 1,600 litres per inhabitant. By agreement with engineers from the water department, the local council reduced this rate to 90 litres per minute, or 800 litres per head. This figure was not actually very high, since farm animals (cows and horses) accounted for 20 to 50 litres per day per head. Drawn up in 1911 and accepted for subsidy in 1914, the project was taken up again in 1922 after an interruption caused by the War, accepted again for subsidy in 1923, revised in 1926 and implemented between 1927 and 1930. This was by no means an isolated case. It is true that the French desire for revenge also in certain cases encouraged the construction of water supply systems, at least near barracks and along main railway lines. Thus on 25 September 1908, the Minister for War asked his colleague in the Ministry for Agriculture to install a water supply at the station of Montreuil-sur-Ille, 'in order to safeguard at times of mobilization the heavy traffic of military trains on the line between Rennes and Saint-Malo'.[46]

Nevertheless, between 1920 and 1940, improvements to and extensions of drinking water supplies were more common than new installations, although these were more numerous in the 'backward' departments of central and western France, particularly in market towns. Thus in 1939, Menetreols-sous-Vatan (Indre) requested a subsidy for the construction of a drinking-trough with a 600 cubic metre supply tank fed by two springs, since 'the town [. . .] currently depends for its water supply on a number of wells which provide an inadequate volume of water in periods of drought. In order to supplement this supply, farmers have to travel long distances in order to obtain water in barrels to meet their domestic and agricultural needs.'[47] The scheme was approved on 9 October 1940 by the Higher Committee for Public Health, but the subsidy was not paid until 1946, after the works had been completed!

Thus the provision of water supply networks in the French heartland proceeded slowly, as positivism gradually influenced people's thinking, supported by the undeniable successes of scientists, technicians and entrepreneurs. Progress was made thanks to thousands of long-forgotten schemes, unlike the large-scale and sometimes ostentatious projects that were common in the major cities. A drinking-trough, a wash-house, a pump, a diversion of a water course, greater use of cast-iron pipes, a few well-located fountains, a connection to the supply network of a neighbouring town:[48] these were the kinds of small-scale but important works that were undertaken and which, as a result of a transformation of attitudes, seemed likely to lead to changes in customs and habits.

9

Water in everyday life

A model from on high

It would appear that membership of a particular social class determined the facilities to which individuals had access and their use of water. Thus for the daughter of a manual worker in Lille in 1920, the middle-class use of water was 'a marvel: you turned on the taps and cold water came out of one and hot water out of the other. There were all sorts of pots and jars on a white washbasin, a bathtub to sit in, eau de Cologne and soft towels. For this little girl who had never before left her own courtyard, it was the last word in luxury.'[1]

From the seventeenth to the beginning of the twentieth centuries, more affluent, more knowledgeable and more 'positivist' people, particularly members of the medical profession, possessed the latest, most advanced sanitary equipment. In the houses of Parisian doctors of the sixteenth and seventeenth centuries, 'the entire sanitary installation comprised nothing more than tiny earthenware bowls, together with a water jug and a copper or porcelain fountain.'[2] In the seventeenth century, peasants in the area around Meaux, with the exception of a few 'cocks of the walk', were bereft even of water jugs, either for washing or for their habitual drink.[3]

In the eighteenth century, following the example of Louis XVI who had had a complete British-style bathroom installed at Versailles, the upper middle classes in particular, and not only in Paris, became interested in acquiring equipment and space dedicated to the pursuit of hygiene and health. Thus at the end of the eighteenth century the court surgeon Caignard, a well-to-do man, had a luxurious residence in Versailles with a cellar, orangeries, winter garden, billiards room . . . and bathroom![4] Another rich court surgeon, Boudet, had an aqueduct built in order to convey water from Reuil to Suresnes and a second one built to supply his house in Maurepas.[5]

In the nineteenth century in Brittany, doctors showed an equally undeniable sense of comfort and made it a point of honour to have, if not WCs, then at least one bathroom.[6] Even before 1914, the parents of future doctors had all the sanitary facilities that we now take for granted. A little later, it was the turn of the parents of future secondary and primary school teachers to install these facilities in their homes. In this way, the social hierarchy was respected.[7]

Sometimes as early as the end of the eighteenth century, but more particularly during the 1830s, middle-class households filtered their water and even installed pumps or aqueducts; this was particularly the case in the provinces. Evidence of this development is provided by the emergence from 1840 onwards in Nevers of plumbers whose job it was to fit sanitary equipment.

However, while the equipment may have been increasingly in evidence, habits had hardly changed. Thus in 1900, 'the idea of immersing oneself in water up to the neck was considered a heathen notion', according to the Comtesse de Pange.[8] An upbringing impregnated with Catholic puritanism forbade the washing of any part of the body other than those that were visible and 'socialized', the only ones considered decent.

Until the middle of the twentieth century, boarders in convent schools were not allowed 'to take a bath unless they were wearing a nightdress. . . . At the end of the nineteenth century [. . .] a nun would hand a dry nightdress to a young boarder and say to her while she got changed: "Raise your eyes to heaven, my child!" in order to prevent her from seeing her own body.'[9]

However, some of the boarders, because of the upbringing that they had had within their families, did come into conflict with the nuns.

In 1918, when I was twelve, I was sent as a boarder to Blois, to Sainte-Geneviève, which has since become Sainte-Marie. I was then in the habit of taking a daily bath, before dinner, during which I was vigorously scrubbed by the strong hand of mother or the maid.

When I became a boarder, we had to dress and undress in the dormitory inside a nightdress. Shame, shame! The washing facilities, which were at one end of the dormitory, consisted of a long plank on which each of us had a bowl measuring a good twenty-five centimetres in diameter and a jug containing a litre of cold water. As a result, we washed like cats: just the tips of our noses and our hands. There was no question any longer of washing our teeth, or anything else, not even for girls who had reached puberty! Those depraved girls who insisted on cleanliness could have a footbath every fortnight (which was charged extra on our bills): we would sit in a circle in the infirmary, each with an oval bowl in front of us in which we were supposed to wash our feet; however, it was forbidden to cross one leg over the other in order to soap our feet as we took them out of the bowl one after the other!

In my very first letter to my mother, I complained about the washing facilities and, during the next prep period, I was called to see Mother Superior who angrily ordered me to rewrite my letter without any mention of hydrotherapy. Although I was flabbergasted, I replied that I would tell her when she came to visit for the first time [. . .].

I then secretly urged the most resourceful of my fellow boarders to persuade their mothers to take some action. As a result, after several mothers had been to visit, five or six were taken under close surveillance to a bath house in town. When we entered our cubicles, linen had been laid out for us. When I got out of the water, I dried myself with one of the two dressing gowns, since I thought that I had been given two gowns by mistake. When matron collected the dry gown with the wet gown, she gave me an icy stare. When we got back to school, I was called to the Mother Superior: Miss G., you have committed an abominable sin (guilty party looks confused!), you only used one dressing gown; you must therefore have bathed naked (guilty party looks even more confused!). You should realize, Miss G., that the Holy Virgin herself has never seen her own naked body. [10]

In fact, until the late nineteenth century, the morning toilet, which had been in retreat since the end of the Middle Ages, was restricted among 'civilized' people, as indeed among many others, to observing the precepts of the school of Salerno:

First wash your hands in clear, fresh water. Splash it on your eyes in order to refresh them. Comb your hair, clean your face and then brush your teeth. [11]

The practice of washing one's whole body, which was slow in gaining widespread social acceptance, even among the elite, reflected sexual taboos and the ethics of sin. Except among prostitutes and dandies, it was considered scandalous to trangress these taboos and moral codes.

The world of the peasantry

Until about 1914, the peasantry accounted for the majority of the French population and remained very attached to their ancient beliefs. To peasants, water was a source of anxiety and of hope, the symbolic representation of the transition from one world to another. It symbolized the hopes and expectations invested in the cults of fertility and health, in which water was endowed with supernatural powers of healing which could be tapped while intoning ritual incantations or performing ritual gestures and perhaps swallowing a few mouthfuls of the precious liquid or washing the injured part of the body.

In the Dijon museum are displayed a series of small sculptures representing heads, arms and legs, which bring home the antiquity of this cult of the springs, which considerably predates Christianity. Similarly, traditional fairy stories, festive rituals and pagan mythology (the word 'peasant' is a derivative of 'pagan') depict water as an ambivalent element. It symbolizes transition, whether from life to death or from death to life. The ancient myth of Charon, the ferryman who brought the dead across the river Styx to Hades, the fairy Vivian in Celtic mythology, the capricious sea that can be as generous as it can angry and murderous, the treacherous water that tempts the living to suicide, personified by the sirens to which Ulysses dared to listen having had himself lashed to the mast of his ship. . . . In this contradictory element, good and evil spirits existed side by side; their good will had to be gained, they had to be tamed and made into allies in order to overthrow one's enemies. The nineteenth-century peasant's perception of water as an element seething with activity was thus little different from that of a much more remote past.

On the eve of the First World War, many peasants still believed that certain springs could exert a real influence on the elements, on the fate and emotions of human beings and on the health and sickness of men and animals; they consulted them as if they were a sort of oracle, and rites and ceremonies were conducted beside them, many of which are unknown to us.[12] The most famous of the Breton pilgrimages takes place not only at the basilica of Sainte-Anne-d'Auray but also at the piscina fed by water from the spring consecrated by the appearances of the saint to Nicolazie in 1623, and the pilgrims never fail to go and drink from the fountain and wash their face and hands. Similarly, in the first years of the pilgrimage to Lourdes, which dates back to 1863, a woman whose child was in a desperate state ran with it in her arms to the grotto where, in accordance with an ancient and very widespread custom, she held it for a quarter of an hour in the glacial spring water.[13]

The elite sits in judgement over the peasant world

Proud of its knowledge and utterly convinced of the certainty of its arguments, the scientific elite busied itself with making a distinction between the healthy and the unhealthy and disseminating its message of hygiene and cleanliness. It wanted to separate wells from dunghills, liquid manure pits and privies, and to remove cemeteries outside parish boundaries. Death, dirt and chaos belonged on one side, life, cleanliness and order on the other. The convictions inspired by the hygienist movement of the period between 1760 and 1860 reached their apotheosis in the *belle époque* with Pasteurism.

The scientific world thus overthrew 'the world order', changed the use of space and of the body and created new objects and new rituals. It supported first the theory of miasma then that of germ; what peasants saw as a supernatural order of which the visible world was a logical reflection and consequence, it recognized as natural phenomena. To that extent, it came into conflict with the world of the peasantry and its ancient customs which considered dirtiness not as a lack of hygiene but rather as a protective barrier, a sort of '*cordon sanitaire*'. Even today, in the *bocage* of Normandy, grey salt is considered 'stronger' than white salt 'because it is washed less'.[14]

In consequence, the enlightened elite described the physique, clothing and housing of the peasantry as the inadequacies and shortcomings characteristic of a class (or perhaps even a race) whom civilization had passed by. Because they did not wash themselves, peasants were placed in the same category as savages, infidels and even animals. They were more than just 'aliens', they were a living death.

The inhabitants of the Dombes, that vast marsh intersected by a few patches of waste ground and dark forests, have pale, livid complexions, dull, downcast eyes, swollen eyelids and lined, wrinkled faces; their shoulders are narrow, their necks elongated, their voices high pitched, their skin is either dry or soaked in debilitating sweat and they walk slowly and painfully [. . .]; they are old at thirty, and broken and decrepit at forty or fifty. They live out their brief, miserable existences on the edges of a tomb [. . .]. Good health is a blessing unknown to them. Born among the sources of insalubrity, they suffer its disastrous influence from an early age [. . .]. They live in a state of permanent ill health, and go to sleep amid suffering only to wake up to their pain. Hardly have the sun's rays penetrated their dwellings than they are trudging along through dank forests to a filthy marsh from which emanates the poisoned gas that they will once again inhale [. . .]. Everything conspires against their health: their dwellings, their habits, their rough, unhealthy, insubstantial food and the indifference with which they choose their drinking water.[15]

'A perfectly internalized social constraint',[16] the bodily hygiene – albeit of a limited nature – that was now widespread among the higher social classes enabled members of those classes to preserve appearances. It provided a perfect ideological cover for dissimulation, sustained as it was by an ancient tradition of cleanliness that went back to the school of Salerno; 'hygiene is uplifting' as Martin Nadaud, a Member of Parliament, said in 1888, since, as he explained, 'I have always noticed that men who keep themselves clean, just like those who are dedicated to their work, are almost all good citizens and family men.'[17]

Although it established a few (albeit provisional) socio-medical truths, this interpretation of peasant culture led observers of the period to paint a vaguely pitying, but ultimately meaningless picture of the squalour of life in the country. They described the clothes worn by peasants as unclean, whereas in the country dirt constituted a protection against disease,[18] and condemned the practice of washing clothes and linen in rivers because they were of 'dubious cleanliness'.

Dirty people were said to be contagious, even harmful, and those engaged in 'pernicious' occupations, such as grape-picking, were right at the bottom of the social scale. This is why, in accordance with the demands of Fourier, popularized in 1834 by Considérant in *Destinée Sociale*, a novelist like Eugène Sue, a descendant of a long line of doctors, had no hesitation in Le Juif Errant (1844) in putting forward as a model, in contrast to the 'muck heaps' described by the social observers, a 'municipal building' equipped with a closet in which, in addition to a wardrobe, there was a table for toilet articles and a large zinc bowl under a tap providing an unlimited supply of water. Such facilities, which had existed since the second half of the eighteenth century, existed at that time only in certain mansions, country residences, prisons, hospitals and brothels.[19]

Peasant mores

Generally speaking (that is, apart from the discontinuities that the seasons or illnesses, for example, might represent), peasants did not restrict themselves to a naturalist concept of water. It is of course true that peasants needed a daily supply of water, for their own consumption and that of their animals, for cooking, for making cider or perry, to keep the stockpot boiling, for washing clothes and, as soon as the floor was no longer simply beaten earth, for scrubbing the flagstones. From a symbolic point of view, however, the carrying of water and its social uses were part of the female world, and men were very often excluded from handling it. Peasant custom forbade washing on Fridays, the day when water reflected the transition from life to death, during Holy Week and even during the 'month of the dead'. Good periods also existed, such as that between 15 August and 8 September. The annual or seasonal wash, which was a periodic occurence like housework and the Easter communion, was not an isolated event, but rather part of a much wider symbolism. Occurring at the turning point between the seasons just as baths were at the frontier between life and death, the grand washes of days gone by were experienced as 'genuine revolutions',[20] in the etymological sense of a complete rotation back to the starting point. This explains why they did not take place very frequently.

Contact between the body and water was also dreaded, as if the myth of Charon was ever present in the peasant mind. Like the laundering of clothes, it formed part of the rhythm of life; complete immersion of the body in water was associated with the major rites of passage, mainly with birth and death. This is why, in 1886, 'le Morvandieu splashes water over his face when he shaves, but the rest of his body has never seen a drop of water since the day he was born.'[21]

This deep-seated distrust was reflected in Minot in the 1960s in the reluctance of local people to install and, particularly, to use bathrooms. Since it represented a complete reversal of old customs, it took sanitary equipment much longer than gas cookers, oil stoves and refrigerators to penetrate peasant homes. The slowness of progress in this respect was undoubtedly due to the persistence of traditional beliefs in the benefits of dirtiness, not only for small children[22] but for people of all ages; as some of the proverbs collected in the last century in the French countryside put it: 'If you want to reach old age, don't take the oil off your skin,' 'dirt nourishes the hair,' 'people who take baths die young.' Nor should it be forgotten that the strong odour given off by the body was an indication of sexual potency: 'The more the ram stinks, the more the ewe loves him.'[23]

In Plovézet, a representative sample of 129 families was selected in order to investigate the technological revolution that had taken place in household comfort; the gap between farming families in the group and the rest of the population was particularly striking with respect to water and sanitation.

The facilities available in individual households clearly highlighted social differences: only half of the houses visited had running water: all the artisans, all the tradesmen, but only thirty-eight per cent of the farmers. Twenty-six houses, almost all of them belonging to farmers, had no water supply at all; a bowl on the table served both for ablutions and for washing vegetables. Most of the other houses had a sink and a simple drain hole: only those houses built after 1950 were fitted with a proper drain for waste water. Unlike in towns and cities, waterheaters were extremely rare among farming families: the owner of a large farm was the first to have one installed in 1952, at the same time as a washbasin and a shower. Sanitary installations revealed the same disparities. Thirteen familes, including twelve farming families, had no sanitation at all. Only two of the farming families had inside toilets.[24]

It was only from 1950 onwards in Plozévet and from 1960 in Minot that the first houses with bathrooms were built; however, in Minot in 1970, in an essay on the subject of 'Your House', one child wrote that in his house (he lived in a new house) the bath was used for storing potatoes.

Until the end of the nineteenth century, the washing of clothes in rural France was an extraordinary undertaking, both because of its cultural resonance and because of the mountains of dirty clothes that had to be washed; as a result, it was not part of the daily round but rather a communal ceremony in which 'the ewer spoke . . . the language of the social hierarchy' and in which 'the great spring and autumn displays, which extended over several hundred square metres of meadows and hedges, were a sign of wellbeing, of wealth, of feminine virtue and of acquisitiveness.'[25]

Like everything else of course, the washing of clothes underwent the social and cultural changes that characterized the nineteenth century. Scientific knowledge of the elements constituting man's environment and their action on the organism changed attitudes towards that environment and towards the 'envelopes' (clothes, houses) with which he

The evolution of the Miele washing machine in the twentieth century. (The invention dates from 1846). (Photograph: © Miele Archives.)

protected his body. The introduction of steam, thanks to Chaptal, and then that of synthetic soda, together with increasing urbanization, eventually led to the old-fashioned practices being considered as something admirable belonging to the past that possessed something of the 'beauty of death'. In middle-class households, washing became the responsibility of servants and was done either in the laundry room or by a professional washerwoman in the local wash house.[26] Indeed, until 1914 and, in some instances, until 1940, it was far from being the case that each dwelling had its own water supply, although it is true that water had become more accessible. Thus, rather like in the country, water for washing clothes had to be fetched, as had the wood that was also required; both tasks were particularly tiresome.

In rural areas in the nineteenth century, water was generally speaking neither plentiful nor of good quality, having regard to the scientific and technical standards of the period. Moreover, there is ample evidence on both these points, notably that emanating from the various health councils. However, these two major shortcoming were not specific to the countryside; they were simply more widespread there, probably because of a culturally determined distrust of water. The suspicion with which water was regarded meant that, in order for it to be beneficial, it had to be integrated into a calendar or a ritual that served as a protective screen for those who dared to exploit its power.

Of course natural conditions meant that the situation varied from region to region. They partly explain why rural areas in northern France, a region of artesian and deep wells, were relatively poorly supplied. On the other hand, there were more springs and expanses of water in eastern France and the Paris region, where it was less difficult to obtain a supply of water. In western Brittany, the location of the springs, which were equally numerous but very small and widely scattered, might have encouraged people to use them, but it did pose problems in towns and cities.[27] There is very often a plentiful supply of water in mountainous regions, and also in humid, luxuriant areas such as the Confolentais, where 'the children do not look as though they have any idea that they live in a region where clear water is freely available.'[28]

In limestone regions such as the Loire valley and the Dry Champagne district, where the water filters deep down into the rock, water was fetched from rivers with the aid of barrels and carts, because water from the Loire was fairly often preferable to that drawn from wells. This was almost certainly because wells had a bad reputation; they were often badly maintained, were sometimes frozen in winter and dry in summer and frequently contaminated by impurities from dunghills and liquid manure pits, to say nothing of the village good-for-nothings.[29]

Change took place only slowly, usually after 1914. Prior to this date, the urban concept of the sanitary function of water emerged only rarely to serve as the inspiration for the modernization of wells[30] and fountains. Thus in Nevers in 1800, only one of the twelve fountains serving the daily needs of the population was 'in order and well maintained at the expense of citizen Saint-Phallet'.[31] Similarly, a hundred years later, the conflict between a farmer in the department of Meuse and a retired city dweller, Major Moreau, a landowner at Burey-en-Vaux, revealed differences between two concepts of cleanliness and dirtiness that were as much cultural as social. On 22 June 1908, the retired soldier wrote to the prefect of the Meuse:

I humbly wish to lodge a complaint about a situation that has become truly intolerable; it concerns a house that we live in for a good part of the year and in which we would live even more if we did not fear that we might at any minute fall victim to an outbreak of disease. Moreover, what I had feared for a long time has now happened: our well is polluted. Beneath our windows, a few metres from our courtyard and twelve metres from our well, there is a dunghill belonging to M. Charles Saleur which is a veritable refuse heap.

The owners of this dunghill, who live opposite us, have no lavatory and all their excrement is put on to the dunghill. . . Every day Madame Charon empties the chamber pots on to the edge of the dunghill, *usually into the gully itself.*[32] It is therefore hardly surprising that our well, which is situated on the edge of the gully inside our kitchen, is contaminated by the filth that rots in the gully.[33]

Charles Saleur's response to this complaint, put on record by the mayor of Burey-en-Vaux, is a crafty answer to the accusation made by Major Moreau, insofar as it accepts the notion of sanitation and then proceeds to use it against the complainant.

In execution of article 12 of the directive of 15 February 1902,[34] Monsieur Charles Saleur, farmer at Burey-en-Vaux, undertakes to make the ground around his dunghill leakproof and to collect the liquid emanating from it in a liquid manure pit.

Since Monsieur Moreau considers that the unsatisfactory state of the dunghills constitutes a public health hazard that might cause an outbreak of disease in the area, I think I should point out that all the dunghills in the village are in the same state as mine and that for a long time Monsieur Moreau had a dunghill in the Rue du Château opposite his own house; his servant, who was the owner of two cows and some pigs, put his dung there and liquid manure ran off it into the gully just as it does from mine. Major Moreau (retd) is now loudly proclaiming the principles of hygiene, but surely it would have been more charitable of him to set us poor ignorant peasants a good example.[35]

However, such complaints, of which there were few before 1914, did not come solely from city dwellers who had moved to the country, at least for part of the year, only to have their notions of health and hygiene offended. They were also made by peasants and reflect a certain degree of hostility between city and country dwellers, a rejection of 'alien' practices and conflict between neighbours, who viewed each other's customs with suspicion.

Against this background of social conflict, the city dwelling landowner could appear to be the accused party. Thus in 1911, Jacquart, a farmer living in the village of Greuilly (Meuse), lodged a complaint against the poor quality of the water from a tank that he used to supply his cattle. The complainant wrote to the prefect as follows:

On 18 August, I write to complain about the inactivity of my landlord, Louis Verdun, of Billy-les Maugiennes, who since last October has refused on several occasions to empty the tank which contains contaminated water; you did not grant me the protection I sought because my letter was written on plain, unstamped paper. Instead of resubmitting my request on stamped paper, I made haste to empty the tank, which contained 100 cubic metres of liquid manure; unfortunately, we were not able to conclude the operation in time to gather the rainwater that fell on 24 and 25 August.

Since that time, the landlord has done nothing. . . He is aware, however, that veterinary surgeons have blamed the bad water for the disease that has struck down three of my horses, one of which has died.[36]

On 2 December 1911, the landlord submitted his defence to the prefect; in his view, the farmer was at fault for not maintaining the tank, which had been repaired and rendered with cement in 1906 at the landlord's expense. Thus in order to settle the argument, the mayor and his secretary went to inspect the tank in question. Their conclusion, which they communicated to the prefect, was that the tank was contaminated by seepage of surface water, since it filled up slightly even in times of drought. The prefect's response was worthy of Courteline: 'There is no point in pursuing Jacquart's complaint any further: article 12 of the law of 15 February 1902 makes no provision for considerations of health and sanitation in respect of animals.'

Modernization

Fuelled by the ancient dread of epidemics, prompted by fear of 'unhealthy emanations' and supported by various branches of knowledge, some ethnographic in nature, others scientific, the arguments used between 1900 and 1910 were repeated incessantly. The social and cultural differences and conflicts emerge in documents when local people speak of their dunghill, their well, their fountains and their lavatories (when they existed. . .), insofar as experts and 'decision makers' paid greater attention to their views after the 1902 law had been enacted. It should not, however, be thought that this law changed popular attitudes at a stroke. It was in fact part of a change of policy, particularly of health policy, that was characterized by state intervention and the outlining of a public health system; it stood in a direct line of succession from the laws on primary and secondary education, on the reform of the medical profession (1892) and on free medical care (1893).

Thus the law of 1902 'set the world to rights', in the narrow as well as the broadest sense, and legitimized the activities of those scientists who managed society; henceforth, notions of solidarity and mutualism left their mark on administrators, hygienists, architects, engineers and political philosophers descended from Saint-Simon.

Once again, administrators, politicians and the medical profession joined forces to bring sanitation to places which epidemic diseases such as typhoid had shown to be contaminated. In July 1903, when asked to bleed six patients during a mini-epidemic then raging at Frontenay-Rohan (Deux-Sèvres), the doctor in charge of epidemics laid the blame for the outbreak on the water from a rural well:

These patients, found within 10 metres of each other, were struck down at the same time; their condition is serious; they all drink water from the same well; their illness must therefore have a common cause. In this case, we are coming to the same conclusions that we reached during the great epidemic at

Champdenier; beside this well there is a slaughterhouse which discharges its waste water and refuse into a cesspool, which is apparently an ordinary well. Because of their close proximity to each other (they are approximately eight metres apart), they are fed by the same underground water supply, which is contaminated by the cesspool.[37]

Consequently, in accordance with the law of 1902, six samples of water were taken on the order of the prefect from the affected parish; analysis carried out by the laboratory of the school of medicine in Poitiers confirmed the doctor's opinion.[38]

Thus in view of the frequency of cases recorded during epidemics and of neighbouring cases, a ministerial circular of 1924 clarified the conditions for the monitoring of water laid down in article 9 of the 1902 act. Water monitoring was henceforth to be the responsibility of the public health inspector in each department. In some departments, this official worked remarkably hard. In the department of Oise, for example, Dr Pacquet drew up a health file for each parish, even though the information collected was often sketchy. He set himself a certain number of priority tasks, such as being stricter for the year 1926 about the number of monitoring samples taken, obtaining the authorization to have samples analysed before a new water source was tapped and continuing to ensure, in collaboration with expert geologists, that regulations governing the protection of the catchment area of wells were properly enforced.[39]

Thus legislation and statutory regulations, local officials and official enquiries – such as the one ordered by the circular of 9 January 1913 for parishes with more than 5,000 inhabitants[40] – combined to improve the monitoring of the water supply and to increase the number of analyses carried out. During the 1900s, circulars to prefects from the Ministry of the Interior, issued by the Department for National Assistance and Public Health, repeatedly stressed:

the overriding importance for public health of the rigorous application of article 9 of the law of 15 February 1902. It is extremely desirable that municipalities, advised and guided by the health organizations and yourself, understand that it is both in their interest and their duty to provide, as the law states, 'a supply of drinking water of good quality and in sufficient quantity', to make provision for the drainage of waste water, to improve the sanitation of the most insalubrious districts and thus to reduce the most important causes of mortality and morbidity, to safeguard productive labour and to work for the good of their community.[41]

Nevertheless, the public health act of 1902 cannot be reduced to a number of measures imposed on the populace from on high. Even before the legislation was enacted – and particularly between 1830 and 1900 – there were many requests and even petitions from municipal councils in market towns and villages seeking the implementation of public health regulations and the cleansing of the water, soil and subsoil. Indeed, faced with the success of Pasteurian medicine, rural communities established links of causality and tended to win acceptance for them.

In 1899, the inhabitants of Laqueille (Puy-de-Dôme) devised a scheme for replacing 'water of dubious quality' with 'very pure water from the Roumières springs'.[42] In Pérignat (Puy-de-Dôme), it was the mayor who noted in consternation that 'it was not uncommon during rainstorms to see toads and all sorts of other little animals gushing forth from the public fountains'.[43] In Longes, a village with 768 inhabitants in the department of the Rhône, subscriptions were raised from the people requiring a water supply (260 in all); the sums paid 'varied from 1 to 800 francs. The poorer members of the community who could not pay in money undertook to pay in kind, that is say they promised to supply their labour on a daily basis during the realization of the scheme.'[44] A total of 170 days' labour were pledged, made up of 159 days of human labour, calculated at 3 francs per day, 11 days of oxen labour, calculated at 8 francs per day and 13 francs' worth of smith's work. In 1886, fear of catastrophes caused by drought and fire facilitated the task of the municipal council in Germangat (Ain);[45] it had no difficulty in gaining acceptance for its view that a water supply justified the imposition of an additional tax.

Enthusiastic workers in the cause of public health expressed their satisfaction with this increasing awareness in reports to the departmental committees for public health (which were passed on to the National Consultative Committee). Thus one doctor in the Aisne department wrote:

As a consequence, I propose that we should express a favourable opinion and congratulate the municipality of Marle on its intelligent initiative.[46]. For more than fifteen years now, hygienists have been stressing to municipalities the importance of water: we have not been preaching in the desert, since today we can see municipalities all over the country vying with each other to provide their inhabitants with a good supply of drinking water.[47]

Not only in towns and cities,[48] but also in rural areas, local authorities were being encouraged in various ways to see improvements in sanitation standards as a solution to the social and health problems that they faced; in this, they had the agreement of a population keen to avoid a repetition of biological misfortune[49] through the intervention of technology in the world order.

Thus the mayor of Jouques (Bouches-du-Rhône, administrative district of Aix) 'made it known to the municipal council that he had been worried for a long time by the state of health of the population, which was continually and at very short intervals decimated by outbreaks of typhoid and paratyphoid. In order to find a solution for this disastrous state of affairs, the mayor had consulted several hygienists who all agreed that the cause of the problem lay in the very dilapidated drinking water pipes that supplied the public system.'[50]

In order to answer the anxieties of local people, municipal authorities established water supply companies, unlike some towns and cities which preferred to grant the concession to a private company, thus giving up some of their powers.

Diversity among the peasantry

The relationship of the rural population to water thus reflected the process of 'hygienization' then in progress, and many peasants, having been duly reprimanded, were themselves receptive to the argument presented to them. However, there is no denying that this sort of relationship was characteristic of only part of the peasantry; its adoption by certain elements did not eliminate other, different relationships.

Two broad types of relationships emerged. The first of these was the 'ethnographic' type, typified by a tenant farmer from the Confolentais region, of whom P. du Marroussem wrote:

The people are dirty, and this family has not lost sight of that tradition. . .
Both men and women are concerned above all with worship of the dead and saying prayers for their livestock and the harvest. . . The furniture and the dwellings do not have the sparkle of Flemish farms. The children do not look as if they suspect that they live in a region where clear water is freely available. On Sundays, however, the family is very properly attired.[51]

According to this observer, progress was being made with respect to cleanliness, but use of the *'langue d'oc'* presented an obstacle to the dissemination of the rules of hygiene. Moreover, 'illnesses caused by humidity, from pleurisy to toothache', were treated as much by 'supernatural agents' as by the three doctors who worked in the region. 'Saints are first and foremost protectors of livestock, and their protection can only be obtained by drinking at the springs that make the meadows fertile.' According to this belief, spring water was an ambivalent element: it was purifying, life giving and regenerative, the inducer of all fertility, both seminal fluid and moist womb; and by virtue of being the first giver of life, it could also give back, prolong and save it.

The very unchristian cult of the saints, both evil and benevolent, shows the importance of water in the rites performed by the healers of the region:

The process is called 'recommendation'. It is conducted in the following manner. The woman who tells fortunes by the saints takes a hazel twig gathered on midsummer's eve and places a piece of charcoal, also made from hazel, on water that she had poured into a glass. Then, muttering prayers unknown to the uninitiated and handed down from generation to generation of sorceresses, she utters the name of a saint. If the charcoal remains motionless, the saint is innocent and she passes to another one. However, if the charcoal sinks to the bottom, the evil saint is known. The only thing left to be done is to appease him by fetching a phial of holy water from one of the many holy places in the area and using it to wash the diseased member, on three consecutive mornings at sunrise. If needs be, if you are embarassed by fear of what people may say, the good lady will undertake the pilgrimage for you.[52]

Thus through agrarian rites typical of farming people and a healing ritual common to many rural populations in France up until the present day, the numerous miraculous springs that still exist in France, from Sainte-Anne-d'Auray to Lourdes, bear witness to the virtues of running water and the fountains of youth. This association between water and fertility is a virtually universal belief and is apparent in all the rituals of regeneration through baths, immersion, sprinkling or ingestion which in both Japan and Europe still form part of hydrotherapy, a practice in which myth and legend have been taken over and strengthened by medical science.

A second type of peasantry, different from but not the exact opposite of the first, is typified by the northern French wine-growers of Ribeauville in Alsace, who were observed in 1888 by Charles Hornell 'a vineyard owner and president of the district agricultural association'.[53] In Ribeauville, just as in the Confolentais region, the cult of the dead, the veneration of the Virgin Mary and the giving of alms were common practice in these two Catholic families. However, the wine growers of Ribeauville belonged to an urban environment and had the mentality of city-dwellers. They were extremely pious, bigoted even, went to mass every Sunday, voted in accordance with their priest's instructions and subscribed to a religious newspaper; they owned their own 3 hectare vineyard, contributed to two relief funds, paid for the services of the doctor, pharmacist and midwife and were sober, honest, provident[54] and therefore in good health! 'The family has acquired the habit of cleanliness. They wash frequently. Each morning and after work that is even slightly dirty, winter and summer alike, the father and his sons go to the nearby well as soon as they arise and wash their face and hands. The wife washes inside the house.'[55] This Alsatian family, which came originally from one of the two lost provinces, is presented as a model to be imitated, and the observer concludes: 'In sum, the family that I have just observed and described is built on religion; respect and love for the mother and father are the foundations of this venerable edifice, order, economy and work its pillars and prosperity and happiness its crown.'[56]

This family of small landowners, which was well integrated into capitalist society and made its daily, albeit limited sacrifices to the modern ritual of hygiene, was endowed with all the virtues, unlike the superstitous, dirty, improvident and therefore unhealthy peasants of the Confolentais region. Between these two very different types – which verge in certain respects on caricatures – there was a multiplicity of other types; some were very similar to one of those described, some very different, while others displayed characteristics of both types. The 'water culture' was sufficiently vast and welcoming for belief in the sacred and in the purifying, regenerative powers of running water to coexist, within the same family or even within the same individual, with beliefs of a more scientific and technical nature, such as the careful selection of drinking water.

Moreover, as in the French West Indies today,[57] the high value placed on purity could constitute a backcloth on to which modern hygienic practices and equipment could be fitted. However, these new practices did not necessarily overthrow the traditional health system prevalent in peasant communities. Just as the American Indians accepted the Spanish *conquista* and the Christianization thrust upon their native religion, peasants in nineteenth-century France – or a large number of them at least – as witness their present-day descendants who have been put under the microscope by anthropologists, retained their old perceptions and behaviour, reinterpreting hygiene in their own way and enlisting it as an additional means of assistance. Although water, a scientific and industrial product, could bring immediate assistance, it could not provide diagnoses, since hygiene, like modern medicine, tends or – possibly – prevents injuries, whereas traditional medicine tended to treat the original source of anxiety. To this extent, modern hygiene and traditional medicine were able, in the eyes of the peasantry, to share these two roles between them and endure side by side in such a way that 'this compatibility resolved the contradictions inherent in their coexistence.'[58]

Water and housing

Just as the eyes are the 'windows of the soul' and the skin represents its casing, the house is a symbolic retreat, and a social mantle, for bodies fatigued by work. An aspiration as much as an obligation, 'finding shelter in a dwelling had become by the end of the nineteenth century a subject of study and consideration for intellectuals, writers, politicians and hygienists. As a consequence, destitution and poverty had become a scandal and housing a social problem.'[59]

Knowledge of agents that modify the environment, such as water, linen, washing and architecture, and of their chemical composition, was henceforth to be harnessed in the service of human comfort and the good of society as a whole. Thus, in the words of Victor Considérant: 'In a communal building, everything is planned for and provided, organized and thought out; here man rules, in perfect control of water, heat and light.' And in 1839, Ch. Harel put forward proposals for a communal building in which each house would have central heating, a bathroom, double walls and, as a result of collective organization, reduced laundry costs.

Until about 1840, the countryside and most towns were entirely devoid of water supply and distribution systems. And in working-class districts, there was no water for cleaning the streets, for houses or for domestic purposes. The cholera epidemics (1832–3, 1849–50, 1854–5, 1865–6, 1873, 1884–5, 1892), together with one national typhoid epidemic, lent impetus to the search for solutions to the problems of water supply and sanitation. However, it was also a question of finance, since houses in urban areas often had wells, while private mansions and middle-class apartment blocks had access to street fountains and filtered their drinking water. And it was not until 1893 that legislation was introduced to force householders and landlords in Paris to have their properties connected, within three years, to the main drainage system.

This legislation was the culmination of the efforts of campaigners from widely differing backgrounds. Some were adherents of the utopian socialism of the 1830s, others of the philanthropic school, while yet others were simply concerned with order and salubriousness.[60] A great mass of documentation – including Villermé's *Tableau* of 1840, some of the publications of Engels[61] and Marx, some official reports – such as the Chadwick Report of 1842 and the one on working-class housing in Lille published in 1843 by the Société de Saint-Vincent-de-Paul – the account of Blanqui's journey to the major industrial cities (notably Rouen, Lille and Lyons), Frégier's symptomatic book on the 'dangerous classes of the population of cities', *L'Extinction du Paupérisme* (1844) by a certain Bonaparte, the translation in 1850 of a paper by the Englishman Henri-Roberts commissioned by Napoleon III, the surveys carried out by Frédéric Le Play into the 'workers of two worlds', to say nothing of descriptions by both great and minor writers[62] – bears witness to a heightened and many-sided sensibility, ranging from a concern with social order as a guarantee of stability to a desire for change and revolution. This latter desire gave rise to a proliferation of descriptions considered by some to be 'literary', even 'miserabilist'. In fact, the elites were once again using their excess time to give expression in writing to their various approaches, but they came up against the housing issue, a very real problem situated at the point of overlap between the cultural, the sanitary and the social.

Shortly before the promulgation on 13 April 1850 of the law on insalubrious dwellings, no doubt in order to prepare what then passed for public opinion, the conservative Second Republic turned its attention towards the question of hygiene in the home, and particularly in working-class houses. Thus on 21, 26 and 27 August 1849, the editor in chief of the *Moniteur Universel* published a report on 'The progress achieved in the matter of insalubrious dwellings', which has the merit of giving clear expression to the Government's intentions:

Charity, philanthropy and economic science all lead us to declare that it is of great importance that the poor, and particularly workers in the towns and cities, should live in healthy dwellings; they all acknowledge the threefold benefit to be gained from this. Firstly, there is the material benefit of the health, vigour and physical wellbeing of individuals; secondly, the moral benefit of a clean and salubrious dwelling, which exerts a strong influence on the spirit of the family and prevents the worker from feeling aversion to a home which he finds repulsive and damaging to his health and from which he is eager to flee in order to seek his often fatal pleasures outside the family home; and finally, there is a national benefit, since it is in the country's interest that its citizens should be vigorous and strong when called upon to defend it. . .'

In fact, the housing policy outlined here was based on a programme of moral order. Physical wellbeing, the preservation of the family and the maintenance of morality were three of the pillars on which the security of society was based. As the place of intimacy, fertility and conjugal duty, the home was invested with a social and cultural value which, by virtue of its improvability, served to condemn 'vices' and 'social scourges' and to glorify their antidotes.

Nevertheless, if the sketch published by the *Moniteur Universel* in 1849 is to be believed, the peasantry enjoyed a healthy life and lived in dwellings which were not overcrowded, unlike manual workers in towns and cities 'where hygiene conditions were bad'. Thus the stereotypes of the peasant and the city dweller, which go back at least as far as Rousseau, served to express the distinction between healthy and unhealthy and to justify the attention brought to bear on the working class.

This was why the law of 13 April 1850, on the express order of the French Government, set up a committee on insanitary housing in all local authority districts that considered it necessary. Probably because it was already aware of the lack of urgency with which mayors and municipal councils were setting up these committees, the Ministry of Agriculture and Trade[63] sent a circular to prefects requesting information on the implementation of the circulars of 11 and 20 August 1850 on the improvement of sanitary conditions in insalubrious dwellings and on working-class housing. In fact, to judge from a sample study carried out in seven departments,[64] there was no doubt about the conclusion: with the exception of the city of Lille,[65] the new legislation did virtually nothing to improve conditions in insanitary dwellings of the period, as Rouher, then Minister of Agriculture, Trade and Public Works, acknowledged in the circular that he sent to prefects on 27 December 1858.[66] Faced with a popular culture which did not necessarily identify cleanliness with hygiene or clean water with physical health, the knowledge of the hygienists and the political concern of the administrators were thwarted. Indeed the latters' intentions were frustrated to such an extent that their attitude swung between the two extremes of zealous activism – which was very rare and could be interpreted today as coercion – and resignation, which was often tinged with sadness and sometimes also with a sort of prudence. As the sub-prefect of Ribérac wrote: 'There have been many improvements, but they are the result of prosperity and the times in which we live; housing conditions are gradually improving and people are eating better food. The improvement in housing conditions will soon become as well established as the practice of eating meat is already.'[67]

The peasant's 'hovel' described by La Bruyère deserved its label, which is unworthy of the man, only because of the development of a refined city-dwelling world which served to highlight 'the scandal of the difference in ways of living'.[68] From the poor man's house – a more or less permanent fixture – to the palace of kings and their successors, housing reflects the language of the social hierarchy and cultural differences. Thus the houses in which the day labourers of Beauvais lived in the seventeenth century were very similar to those described by nineteenth-century folklorists and social commentators: 'Quite a number of day labourers own their own house, a modest one-roomed cottage with an attic in the roof and flanked by a stable, a small barn and a garden of several hundred square metres. The interior is furnished with a few crude items of furniture, straw mattresses, clay crockery, two or three pairs of sheets, a few hemp shirts and clothes and a blanket made of serge.'[69] Such cottages contrast, for example, with a parish priest's residence and outbuildings in eighteenth-century Authie: water was supplied from a well and two lavatories provided what we would consider as basic hygiene. This priest's residence gives us some idea of what the houses of well-to-do peasant farmers must have been like in the eighteenth and nineteenth centuries.[70]

As far as towns and cities were concerned, we know from Nicolas Delamare's *Traité de Police* of 1722 and from recent theses on the seventeenth and eighteenth centuries[71] that they had not really solved the problem of disposing of waste water, although water supply systems tended to be much more adequate.

Under these circumstances, the 'poor man's dwelling' continued to preoccupy municipal officials and councillors. In 1890, O. de Mesnil, in his description of the general state of working-class dwellings in Paris, noted that they were cramped, not connected at all to the sewer network and lacking in any system of refuse collection. Moreover, he pointed out that in many dwellings 'water was entirely absent.'[72] On a visit in 1882 to the avenue de Choisy, he reported what he had seen: 'A house with a mud floor, covered in tarred felt and in poor repair, the windows of which have no glass and are blocked up with scraps of muslin. The tenant sleeps on a bed of wood shavings strewn on the floor and pays 12 francs per month, paid in advance. This house has no chimney, no water and no privy.'

On the Boulevard de la Gare there was a two-storey apartment block connected to a drain into which only rain and household waste water flowed. The author concluded that they were 'veritable cellars', and added that some of these dwellings were 'in such a filthy state that they are no longer inhabited and are gradually becoming rubbish tips'. 'Since they have neither water nor toilets, they are disgustingly filthy.' The same was true of the rue du Château-du-Rentier, where pigs were kept in the bedrooms. . .

Obviously, in Paris in 1882, just as in the eighteenth century, the new public health legislation was not obeyed, and the sight of hovels that looked like those seen in the countryside was a profound shock to the observer of the 13th *arrondissement* as it had been to the observer of the squalid courtyards of cities in northern France.[73]

The statute of 25 October 1883 on the renting of furnished apartments in Paris amounted to little more than wishful thinking when it declared: 'There should be at least one lavatory for every twenty residents. These lavatories should be equipped with flushing mechanisms and automatic water seals [Art. 18]. Every house let in furnished apartments shall be supplied with a quantity of water sufficient to keep the building clean and conducive to health and to meet the needs of the tenants [Art. 22].'

However, a major difference between this legislation and the 1850 law on insanitary dwellings was that a team of inspectors, architects and doctors appointed by the City of Paris was set up; between July 1883 and the end of 1884, they inspected 8,000 buildings let in whole or in part in furnished apartments. The law of 15 February 1902 marked a further step forward; under the terms of this act, mayors were bound to receive complaints from individuals, to pass them on to the prefecture of their department and, after an inquiry, to remove the cause of complaint. Nevertheless, it was not until 1940 that virtually all apartment blocks in Paris (97.6 per cent) had running water and mains drainage (94.1 per cent).[74].

The Siegfried bill on council housing, tabled in 1892, was passed by the Senate in 1894. The bill's promoter was delighted to see architects beginning to be less contemptuous of such important questions as 'drain traps, automatic valves, inspection holes, discharge pipes and washbasins'. Moreover, the law also decreed that council houses should have a water supply for the kitchen sink and in the lavatory.[75]

In 1895, an international hygiene exhibition was held; at the 1889 exhibition, one of the displays was 'a comparison between the sanitary and the insanitary house';[76] in 1900, the display mounted by the City of Paris included a 'health file' on all the houses in the capital; for its part, the Housing Commission announced that it was gradually ridding the capital of its insanitary dwellings.[177] After the defeat sustained by France in the Franco-Prussian war of 1870, just as after the Great War, public health, as reflected in sanitation policy, was considered to be vital to the nation's recovery; thus the provision of allotments, council housing, washrooms, shower baths was, like gymnastics clubs, seen as a patriotic act. In 1896, the benefits of loans for 'cheap' houses were extended to shower baths and allotments; similarly, in 1925, the Loucher act encouraged home ownership in the suburban developments that were springing up, for example, around the old gates of Paris.

Despite the failure of the 1848 law on public wash houses, public wash-houses were successfully established in Paris (1867) and in some small towns. After 1870, an association of laundrymen and wash-house owners was set up, and the number of private wash-houses began to increase in major cities such as Paris and Bordeaux.[78]

By about 1900, whether or not it was served by wash houses, public toilets, municipal shower baths and, much later, swimming pools, 'the poor man's dwelling' was urbanized for a small minority of the working classes; by about 1920–30, the proportion thus housed had increased significantly. At the same time, real changes were taking place in isolated farms, hamlets, villages and market towns, albeit at a very uneven pace and often slowly enough to pass almost unnoticed.[79]

Finally, the rural exodus, which did not really gather pace until after the First World War, meant that many peasants settled in towns and adopted urban habits. Thus the journeyman carpenter, a socialist married in a civil ceremony without benefit of clergy, who lived on the edge of the Saint-Germain district close to the Ecole Militaire in a building occupied by a girls' school 'had adopted the habits of the middle classes. The family had come to respect the medical profession and old wives' remedies were considered to be nothing more than superstitions.'[80] For both partners, one born in the department of Indre, the other in the Dordogne, both of whom had settled in Paris, although without any savings at their disposal, cleanliness was a 'duty', as was love of work and filial devotion.

The same was true of a former seminarian, working as a printer, who was observed in Paris in June 1861. He had a significantly higher standard of education, if only because of his trade, and supported 'the maintenance of law and order during the events of 1848–1850–1852.'[81] Born in Tours of Catholic parents, married for a second time to a girl from Saint-Julien (Lot-et-Garonne), 'he has never, thanks to his temperament, had any serious illnesses. Moreover, he seeks to prevent them through good hygiene, and when illness threatens, he allays the symptoms by resting and taking special care. His dwelling is always clean and well aired . . . It is often towards his family that B. . . is most solicitous. . . He generally makes his children take a bath once a week in summer and once a month in winter. The doctor is called only in serious cases and he has never disapproved of the initial care given by the father.'[82] He lived in well-kept lodgings, always wore clean clothes and had 'a veritable propensity to save, which is quite rare in his trade'.

Whatever their political opinions, the worker married without benefit of clergy and the former seminarian shared the same relationship to water and the same concern for cleanliness and health. Cleanliness and the washing of body, clothes and lodging were perceived as a method of protection, with the doctor as the final defence against the extreme danger that illness represented. These few individual cases are evidence of the emergence within society of a sharp division of opinion as to the value to be attached to the appearance of the body and of whatever came into contact with or covered it. Among the more traditional elements of the peasantry, who constituted a veritable repository of ancient customs, dirtiness was considered to be a protection and to give greater strength to the healers. Moreover, washing was often held to be unlucky, perhaps because it was seen as a return to the original state of disorder. On the other hand, to those classes influenced by public education and medical science, dirtiness was anathema, the source of all evil and a sign of every conceivable vice. Since the concept of time as effort manifested in the notion of saving (of the body, of semen, of labour and of money) was replacing that of the immanence of time, the relationship with water was changed, if not completely transformed, to the extent that the meaning attached to it tended to be reversed, with all the disturbances that this reversal caused in individuals, between generations and among social groups.

Water and work

Whether it was among the working classes, the urban middle classes or in factories, the relationship to water changed only very slowly, even after 1900; moreover, a great deal of encouragement was required. In 1894, legislation was necessary in order to force the owners of apartment blocks in Paris to connect their properties to the mains drainage system.

The flushing toilet, the use of which had progressed rapidly in Britain at the end of the previous century, was accused of wasting water.[83] The chamber pot, the contents of which were simply thrown into the street, and the close-stool so beloved of Voltaire reigned supreme for a long time.[84] The comic power of farts, turds and piss, which was still very widespread in France in the eighteenth century, did not diminish at the end of the *ancien régime* as quickly as hygienists, moralists and . . . those embarassed by it would have liked. The 'art of shitting', particularly in the open air, was still flourishing at the beginning of the nineteenth century, in mockery of the porcelain of the water closet.

'Aerism', particularly for delicate souls and those with sensitive noses, then hygienism, which treated the reproductive and excretory functions with the same scientific discretion, and finally Pasteurism, which asceptisized the scatological humour of our ancestors, all helped to drive away miasmas, germs and microbes and to oust them from water, which was henceforth purified.

Thus the 'aerism' of the 1780s was still omnipresent both in the questionnaire and in the responses to the survey 'on agricultural and industrial work' conducted in accordance with the decree of 25 May 1848.[85] The survey actually provides little useful information on the cleanliness of manual workers in 1848, since it was the employers and members of industrial tribunals who filled in the questionnaire. In their view, while work tired the body (which was, by implication, quite normal), workplaces and dwellings were not in themselves 'insanitary'!

Thus for carpenters in the construction industry, 'working on the shop floor is more tiring than working on building sites; there is nothing dangerous about it which could be injurious to health.'[86] Similarly, the Printers' Association of Paris replied in 1848: 'There is nothing in our industry which is unhealthy in itself [sic!]. It is sufficient therefore to keep the workshops in a suitable state of cleanliness, provide adequate ventilation, etc.'[87] In the Parisian earthenware, pottery and porcelain industries, the factories were, according to the replies, large, well ventilated and 'in no way injurious to health'. Only rarely did the industrial tribunal stress housing conditions; Angilant, a master cobbler, was an exception when he wrote:

Most of our manual workers take furnished rooms when they are bachelors. In order to reduce the costs of their lodgings, several of them rent a room together. These lodgings are generally rather badly kept and give off a quite disagreeable odour. . . When they get married and find unfurnished accommodation, they keep themselves fairly clean as long as there are no children. However, as soon as children come along and the same room often houses the father, mother and 2 or 3 children, you can forget about the cleanliness of the household.[88]

A quarter of a century later, the parliamentary inquiry into working conditions in France,[89] to which only employers responded, bears witness to an equally obvious lack of awareness of the problems of hygiene and cleanliness (and health) which were part of the daily life of many workers. According to the employers, hygiene arrangements in factories were satisfactory, industrial work had a beneficial effect on workers' health, housing conditions were improving significantly and provident schemes were on the increase, as a result of compulsory deductions from wages.

Once again, the voice of the ordinary worker is not heard. As far as industrial work is concerned, the employers say that it 'develops dexterity and intelligence and trains the eye.'[90] Its effects on health are ignored, as is the question of sanitary equipment in workshops and factories: there are no observations or descriptions, just a few expressions of self-satisfaction on the part of employers.

In fact, it was not until the laws of 17 June 1883 and 1 July 1903 and the decree of 29 November 1904 that a specific body of legislation was introduced to control hygiene conditions in industrial establishments. However, even as late as 1921, fourteen years after the establishment of the Factory Inspectorate, this legislation was not always satisfactorily implemented. Thus in the artificial flower industry in Paris, 'the employers have made many improvements. Although it is not yet possible to use totally harmless colours, at least it is easy to use alcohols and glues of good quality and to forbid the use of bleach by giving workers time to wash their hands properly in hot water. However, it must be said that there are uncaring and negligent employers.'[91]

In the publishing industry, and more particularly in printing shops, where lead poisoning was still a hazard, the situation was no better;[92] thus in 1921, the second Publishing Conference passed a resolution 'demanding' the implementation of existing legislation.

In the same year, the Ministry of Labour sent the text of this resolution to all divisional heads in the factory inspectorate. A survey of all printing works was carried out with a view to improving the 'hygiene situation' on the shop floor.

However, this was hardly an innovation on the part of the Ministry of Labour. In 1911, under the leadership of Joseph Paul-Boncour, it had carried out another 'survey of hygiene and sanitary conditions in typesetters'.[93] Apparently, this survey had not provided the results that had been anticipated, so that even after the First World War monitoring of hygiene and sanitary conditions in printing works was still a goal to be pursued. Indeed in 1922, the Minister of Labour wrote: 'I am not unaware, and worker representative organizations in the industry do not hesitate to acknowledge the fact, that workers do not always attach to individual hygiene and that of the workplace as a whole the importance that is required in an occupation exposed, as theirs is, to particular risks from insanitary conditions. Workers need educating in this matter as much as employers do, and responsibility for this education lies with the Factory Inspectorate. . . .' Thus the 1921 survey, like that of 1911, was intended both to reveal the existing situation and to improve it. Analysis of the results obtained in eleven departments and a total of 184 printing works shows that, as far as individual hygiene was concerned, the standards governing material conditions (communal wash rooms, supply of running water, direct drainage of waste water, provision of towels and soap) were basically complied with, although there were a few exceptions.

In fact, as the reports of the divisional inspectors show, even in an industry in which the risks had been acknowledged for a century and in a profession made up of one of the elites of the working classes, the existing installations left something to be desired, to say nothing of the use that was made of them. Thus the inspector for the Paris district, after a survey of 884 establishments carried out by his department, wrote of the wash rooms : 'They are sometimes mere water points, which means that it is impossible for a large number of workers to wash themselves properly and quickly. These water points are very often nothing more than a tap in a courtyard.'[94] As far as towels and soap were concerned, he noted that they were supplied to the work force in only half of the establishments surveyed. Although the questionnaire did not mention WCs, some of the inspectors in the Paris district did point out that 'on this point, there are often grounds to bring forcibly to the attention of employers' the fact that the WCs 'leave something to be desired, both with respect to cleanliness and to their location relative to the work shops.' Reports by divisional inspectors in Dijon, Lille, Rouen, Bordeaux, Toulouse, Lyons, Tours and Strasbourg all revealed the same situation, albeit with a few local variations. Sanitary equipment was thus scanty, comprising a small number of wash rooms and characterized by the absence of towels and soap.

The monitoring carried out by the factory inspectorate of the time was something of an illusion, even after the enrolling of assistants from the working classes, who were disinclined to collaborate with the employers. In 1908, there were 128 inspectors . . . and a total of 512,000 establishments to monitor![95]

Even in printing works, the use of water and its auxiliaries (soap, towels) does not seem to have been widespread in 1922, although article 12 of the law of 25 October 1919 had extended the law of 9 April 1898 on accidents at work to occupational diseases, and in particular 'all cases of lead poisoning of occupational origin',[96] the extent of which had been revealed by a survey carried out in Paris hospitals in 1911.[97]

However, the efforts of the inspectors were concentrated not on individual but rather on collective hygiene, that is hygiene in the workshop. Fear of tuberculosis, which manifested itself in dust, particularly in the textile industries, gave rise to a decree issued by the Ministry of Labour on 10 March 1894, article 1 of which stated that: 'The floor should be cleaned thoroughly at least once a day before and after work. Cleaning should be effected by washing or with the aid of brushes or damp cloths if conditions in the industry or the nature of the floor covering make washing impractical.' This decree was probably not widely obeyed, since it was reissued by the Ministry of Labour on 10 July 1913 (article 1, § 2 and 3).

Thus, in response to a complaint lodged by the Union of Employers' Associations in the French textiles industry, the divisional inspector for Lille wrote to the Ministry: 'It (i.e. the regulation in question) is much too important from the point of view of hygiene for inspectors simply to ignore. Indeed, we are concerned here not solely with inert dust, but also with types of dust that carry along with them not only all the germs in workshops but also those exhaled by people suffering from pulmonary tuberculosis each time they spit, cough, sneeze or even talk.'[98]

However, these two decrees did give rise to numerous writs issued against weaving sheds on the instigation of the factory inspectors: in 1909 at Tourcoing (E.P. and Ch. Toulemonde & Co.), in 1911 in Roubaix (Charles Huet & Co.) and in 1912 (Dubar-Delespaul & Co. and Lemaire-Dilliès & Co.), as well as in 1913 (Ernoult-Dubois & Co.), etc. Even after occupational diseases had been declared notifiable diseases in 1921,[99] the factory inspectorate, which was the subject of constant disputes between unions and employers, at least in large firms, seemed to experience the greatest difficulty in enforcing standards of microbiological hygiene. The problems inherent in a tradition of individual and collective hygiene that was alien to the worlds of both employers and workers were compounded by social conflicts and a divergence of interests.

However, as a result of radical Government plans for public health, reflected in legislation and decrees, and the watchwords adopted by trade unions,[100] health became a priority objective even before 1914, more particularly so in the 1920s and 1930s; suddenly, there was a change in the relationship between water and a working class which was educating itself, becoming urbanized and acquiring faith in the importance of science.

However, this new relationship with water was slow to emerge, since legislation on health and sanitation was particularly difficult to implement. This is reflected in the words of senator Charles Rioux during the parliamentary sitting of 9 February 1909:

A few days ago, in a newspaper an extract from which I am holding in my hand, the following news item was reported: a factory inspector, having found an ill-kept toilet in a factory – which might have been the fault of a worker – issued the owner not with one single summons but with a total of sixty-five, that is one for each member of the work force. The justice of the peace of Saint-Omer decided that the 65 summonses should be reduced to a single one . . . The newspaper adds that, on the orders of the Minister of Labour, the case was brought before the Supreme Court of Appeal, which decided in favour of the Minister, who is sitting opposite me.[101].

And indeed, article 7 of the 1893 law on hygiene and safety at work stated: 'The fine shall be imposed as many times as there are contraventions.' In response to M. Rioux's question, the Minister of Labour defended the factory inspector in question, 'who is known for his moderation'. In fact it was only after four successive visits (between December 1907 and February 1908) and four warnings that 'the factory inspector found it sadly necessary to issue a summons listing 65 contraventions.'[102]

Sanitary conditions in industrial establishments were thus hardly satisfactory. In 1920, 'the director of the municipal hygiene laboratory in Paris examined the effluent from 1,234 septic tanks on behalf of the department of the Seine; he did not find a single one which fulfilled the conditions required by the prefectoral decree of 1910.'[103] The correspondence between individuals and the committees for hygiene and safety gives a similar impression: thus Guinot, the owner of various buildings in Passy, rented lodgings to workers and refused to carry out any repairs at all. 'The WCs in the back yard have been without any flushing mechanism for more than six months, The rusty cistern is full of papers and magazines. These WCs are so badly maintained that it is sometimes impossible to get into them because of the excrement. . . Water from a sink on the first floor is discharged into a simple zinc drain which flows into a gutter running the length of my garage.'[104] As is well known, furnished lodgings in Paris had no modern comforts and the tenants lived in overcrowded conditions, often in a state of penury.[105]

Several newspapers echoed the complaints made by some workers, for example in Paris in 1910. In its edition of 5 April 1910, *L'Epicier Libre*, a monthly corporative publication, named six employers who were not, in the view of the journal, applying hygiene standards in the lodgings that they were letting to their workers. Thus the Maison Durotoy at 131 avenue Victor-Hugo could scarcely be recommended, according to the hygiene department, 'for lodging employees'. The 'rooms are never cleaned, the sheets are changed only every two months and the employees are forced to use their own clothes as blankets, since half of the beds have no covers at all.' In the Maison Damien at 36 avenue de

Beauté in the Parc Saint-Maur, the room 'above the stables is absolutely disgusting'. 'There is one washbasin between six employees' complained the journalist, in support of the workers: he was obviously unaware that this was the legal norm!

On receipt of these complaints, factory inspectors went to the addresses mentioned and in most cases confirmed that the complaints were indeed well founded, even if the real motive behind them had nothing to do with cleanliness or hygiene. Thus in May 1910, a factory inspector from the department of the Seine visited the lodging house kept at 131 avenue Victor Hugo by M. Durotoy, a wholesale grocer, for his employees, and concluded in his report: 'In short, the premises are wholly lacking in cleanliness. M. Durotoy is going to engage the services of a charwoman in order keep the lodging house clean and tidy.'[106]

In fact, sanitary conditions were not always as dreadful as some of the complaints tended to suggest. Thus 'the houses kept by M. Lair, grocer, at 13 rue Masson and 11 rue de Saint-Pétersbourg' offered suitable lodgings for his staff: they were well ventilated, had the statutory airspace per person, were looked after by a charwoman, the beds and floors were clean and the bedclothes were fumigated once a year. 'For the purposes of bodily hygiene, washbasins in enamelled cast iron and with running water have been installed. Dirty linen is stored in a room on the ground floor. The only fault that can be found with this remarkable arrangement is the smell given off by the WCs on the staircase.'[107]

The factory inspectors' requirements were not always accepted by employers; implementation was often deferred, because of the additional costs involved. Sanitary conditions were in fact worse in certain industries than in others; some of the worst were the sawmills in the Landes and the brick works of northern France and the Parisian suburbs, which employed temporary workers who were housed like cattle in worse conditions than those in which many horses were kept!

All the same, some industrialists were able to justify the absence of sanitary installations; thus M. Lebrun, the owner of the Caudos mechanical sawmill 'could not see the necessity of installing lavatories and washrooms, since they would never be used, since the sawmill was a long way from any dwellings and surrounded by fields, brushwood and pine forests and since he did not believe that he could force his employees to use the lavatories and urinals.'[108] The same was true of washing facilities, which were basic to say the least; there were no washbasins and 'at the moment, employees have to go out into the yard in order to wash; water is available only in the tanks used to feed the steam generators.'[109] Clearly, water was considered primarily as a product for industrial use which could be used incidentally and parsimoniously for hygiene purposes, which were restricted still further, particularly in winter, by workers of rural origin who feared the power of water.

Armaments and munitions factories, which for obvious security reasons were subject to greater scrutiny during the First World War, were scarcely any better provided for from the point of view of industrial hygiene. Thus Dr Etienne Marcel and civil engineer Marcel Frois wrote in their report of 15 February 1917:

On several occasions we noticed, particularly in the loading shops of armaments factories, that the provisions of articles 1 (cleaning of the floor, walls and ceilings), 4 (lavatories), 5 (ventilation and heating of the workplace) and 8 (individual cleanliness, cloakrooms with washbasins) of the decree of 10 July 1913 were completely ignored. In almost every establishment, the cloakrooms and washbasins were badly installed or inadequate for the number of employees using them. . . We are firmly of the view that it would be easy, with makeshift facilities, to satisfy the requirements of hygiene. Shower baths are also necessary in all establishments where toxic materials and powdered substances are handled; more widespread use of such facilities would be of great benefit.[110]

Water and washing

Whether it was at the workplace or in the home, so long as there was no supply of running water on the ground floor of every apartment block and no water point on each floor, so long as the use of water for individual hygiene reamined contrary to inherited customs, so long as the absence of heating was an obstacle to any desire to wash during the winter, as long as water remained scarce and expensive and until the threat of cholera brought hygiene into fashion, French people seldom washed. At best, they took the occasional foot bath for medical reasons and, more as a symbolic gesture than anything else, washed their hands and the tips of their noses.

Washing with water was sufficiently uncommon for Michelet to comment in 1867 in his *Journal* on the extraordinary number of bowls of water that his beloved Athénaïs used every day:

She had got washed and dressed very early in the morning, between 5 and 7 o'clock. No woman uses so much water. I admired her. She is always the same, even on days when she is in a hurry. Endless washing.[111]

Washing on a daily basis was still an uncommon habit, even among the middle classes. Thus Edmée Renaudin relates how circumscribed the daily toilet was, particularly in the evening:

A small jug of hot water was brought. 'What shall we wash today?' – 'Well,' replied our Alsatian maid hesitantly, 'your face, your neck? – Your neck? Ah, no, you washed your neck yesterday! – Well then, your arms, up to your elbows, and make sure you roll up your sleeves!' Personal hygiene was attended to while squatting over a bowl. We took it in turns on alternate days. . . .[112]

One is even entitled to wonder what the purpose of the rooms known as bathrooms actually was, when they existed at all, if it was not to put on a display of luxury and power. At the chateau of Aigues, described by Balzac in *Les Paysans*:

the bathroom is lined with Sèvres bricks painted in monochrome, the floor is covered with mosaics and the bath is made of marble. An alcove, concealed by a picture painted on copper which can be lifted by means of a counterweight, contains a bed made of gilded wood in the most extravagant Pompadour style. The paintings on the bricks are copies of drawings by Boucher.[113]

Nevertheless, in both chateaux and middle-class apartments, bathrooms for a long time lacked their most essential ingredient, namely running water. Thus in 1914 in Jerphanion's apartment in the Boulevard Saint-Germain, near Place Maubert: 'The bathroom was not very spacious and had no running water. With the greatest possible ingenuity, the young couple had found place in it for a washstand, a bathtub, and a small shower with a collar and two buckets.'[114]

In the course of the nineteenth century, Britain had become a prudish, puritanical country which was also slowly getting used to the now customary practice of taking baths and showers.

A manual of etiquette dated 1782 advises wiping the face every morning with white linen, but warns that it is not so good to wash it in water, for that makes the face too sensitive to cold and sunburn. A doctor writing in 1801 remarks that most men resident in London and many ladies, though accustomed to washing their hands and faces daily, neglect washing their bodies from year to year.[115]

Habits changed rapidly between 1800 and 1830 in the upper strata of society:

In 1812 the Common Council turned down a request from the Lord Mayor of London for a mere shower-bath in the Mansion House 'in as much as the want thereof has never been complained of; if he wanted one, he might provide a temporary one at his own expense'.

Twenty years later, 'they had agreed not only to a bath, but to some sort of hot water supply, perhaps because of the examples of the Duke of Wellington (who took a cold bath daily), and of Lord John Russell, later Prime Minister, who had designed for himself a great mahogany bath lined with sheet lead that weighed a ton.'[116]

When Queen Victoria came to the throne in 1837, there was not a single bathroom in the whole of Buckingham Palace. It is true that Parliament hastily agreed to make available extra funds for washing facilities, but these amounted only to a bath installed in the Queen's bedroom.

Before the reservations expressed by many doctors and lay people could be overcome, widespread support had to be gained for the school of thought of German origin, which, throughout Western Europe, promoted hydrotherapy in the context of a return to nature.[117] Until about 1930, however, the much more economical and allegedly hygienic practice of providing showers for ordinary people in collective establishments, whether private or public, carried the day.

Even in the United States bathtubs were still a luxury in the nineteenth century. According to a survey carried out in that vast country in 1880, five out of six Americans had no other means of washing themselves than a bucket and a sponge.[118] At the same period and in the same country, not only did workers' lodgings lack bathrooms, they were completely devoid of any sanitary facilities at all. A well-informed author even wrote in 1895 that 'baths with hot water are only rarely used.'

It was about 1900 that the well-to-do classes in Western Europe and North America began to install baths with running hot and cold water in their bathrooms, which generally adjoined the bedroom. Rooms were now more specialized in their uses, and much technical progress had been made, as was reflected in the installation of water supply and drainage systems and water heaters. Large hotels in America and on the Côte d'Azur in France caused a sensation from 1870 onwards by installing running water in almost all bedrooms; it gushed forth from the taps not through the working of a hand pump (as had been the case in about 1850) but as a result of the pressure generated by the precious liquid itself.

Not long afterwards (1920–30), the United States embarked enthusiastically on the production and distribution of standardized sanitary equipment. However, it was from Britain that the 'luxury' bathroom, the forerunner of the more modest ones that we use today, had emanated in about 1800. Installed in hotels in North America, it aroused the admiration of a French traveller who, in 1905, described the Waldorf-Astoria in New York:

Built entirely of red brick in a massive, solid style, right in the middle of Fifth Avenue, it occupies the whole block between 33rd and 34th Street. It cost almost forty million to build. It has seventeen storeys and 1,500 bedrooms, 1,200 of which have a bathroom. . .

The bathrooms, which have tiled walls and mosaic floors covered in wool carpets, comprise (of course) an enormous earthenware bathtub with as much hot and cold water as you want, at any hour of the day or night, a washbasin with hot and cold water and a hygienic water closet. There are a dozen towels hanging on the wall, an electric heater for the curling tongs and ready-printed lists for linen in need of laundering, which can be returned on the same day.[119]

Among the servant classes, as among the mass of workers throughout

France, cleanliness was very limited, both because of unfavourable housing and working conditions, but also because of traditions and social prejudices which prevented maids in Paris from having access to bathrooms. 'Well-kept' dwellings, that is those which conformed to the standards prevailing among the upper classes, remained the prerogative of a minority. The need to fetch water from a fountain – just as in the country – encouraged the economical use of water and washing was often less than thorough.

Victor Gelu, a baker's assistant, used only one 10 litre jug of water for his ablutions for an entire week, despite the fact that his occupation required him to be 'clean'.[120] Fetching water from the fountain was almost always a woman's job; it was an arduous, unceasing task, particularly for those city-dwellers on the upper floors of apartment blocks. Nevertheless, going to the fountain had its lighter side: like the wash-house, it was a gathering place for the village or district, where girls and gossips with high-pitched voices would meet to make fun of their neighbours and mock bashful lovers.

For a long time, even the better-equipped apartment blocks only had a single WC, which was in fact a sort of hut, as in houses in the country, but made of bricks rather than planks and provided with a water point and a cesspit which was, in theory, leakproof. Thus in order to avoid unnecessary journeys, a so-called hygienic bucket was used during the day and an earthenware chamber pot at night.[121]

These buckets, which were collected in the morning and, in theory, returned immediately, were left scattered about the apartment block morning and night, depending on how comings and goings and conversations ensued. In order to avoid this arduous task, many housewives, even as late as 1950 and in towns such as Saumur, Toulon, Marseilles and Beauvais, had no hesitation in emptying their waste out of the window, taking care first to shout a warning to passers-by. The natural slope of the streets carried everything to the nearest stream; otherwise, waste stagnated in those foul and nauseous pools described with horror by the hygienists of the nineteenth century. Lacking water and washbasins, equipped only with a sink which did not always drain properly, these workers' dwellings suffered from smells and 'noxious' miasmas, some of which were caused by blocked drainage pipes, while the water in the wells or water tanks was suspect and often polluted by factories in the neighbourhood; such dwellings reflected a relationship with water that was close to the old type and one which was most certainly as injurious to the health of workers as it was to that of the peasantry.

Water and health

The healing powers of water

Since time immemorial, water, an object of belief as much as of scientific knowledge, has been invested with the 'power of healing'. 'Innumerable pilgrimages, whether collective or individual, public or secret, are witness to the vitality of this ancient belief at the beginning of the twentieth century.'[122] The powers attributed to many fountains were the result of a process of assimilation between the name of the saint who presided over them and that of the infirmity or illness for which relief or cure was to be sought. Thus, for example, in Lower Brittany fountains dedicated to Sainte-Claire or Notre Dame de la Clarté were said to be effective against eye disorders. On a different level, those suffering from boils (known as '*clous*' in French) sought relief at fountains dedicated to Saint-Clou(d).

Other fountains were said to have particular powers as a result of deeds that were supposed to have taken place beside them, or in the locality. A spring at Saint-Géréon (Loire Inférieure) had had healing properties since the Virgin Mary had dipped her finger into it; the fountain of Saint-Méen (Ile-et-Vilaine) was said to offer relief for skin disorders, because the saint had washed in it and had been cured of a skin disease from which he had been suffering.

When the illness was not positively diagnosed and there was some doubt as to which fountain should be chosen, advice was sought in the Limousin region from widowed matrons, and in other regions from old men or witches. 'In Solignac, old women burn a branch of spindle-tree or of hazel tree while reciting a litany of the names of saints to whom fountains in the area are dedicated; when the flame goes out or the branch is burnt up, the saint whose name is being pronounced at that moment is the one to whose shrine and fountain a pilgrimage might usefully be made.'[123]

At Vitrac-Saint-Vincent (Charente), 'the church and the fountain over which it is built are dedicated to Saint-Maixent.'[124] At the end of the nineteenth and beginning of the twentieth centuries, the fountain was said to cure epilepsy and skin diseases. Children suffering from convulsions were taken there and were cured by the miraculous fountain 'even when the doctors had said their cases were hopeless'.

The day of the devotion and pilgrimage was 24 June, when water was sold. In 1945, a Madame Forestier was in charge of the devotions. 'She proceeds as follows: she fetches water from the fountain and washes her hands in it on three successive mornings, then says a prayer and requests that the sick person be cured. It is usually a baby, and she soaks a bonnet or a little shirt in the water so that the healing will be rapid.'[125]

Instead of the sick person going to the fountain in person, it was possible to ask a pilgrim to go and perform the necessary rites on his or her behalf. It was also possible to use holy water to cure dysentry, to reduce a fracture, to encourage successful childbirth or to bring forth the signs of life that made it possible to baptise a newly born infant.[126] In the Charentais region, it was considered until Pasteur's discoveries to be the only remedy for rabies and it also afforded protection against lightning. At Saint-Maurice-des-Lions in 1943 and at Brigneuil in 1944, a lot of peasants sprinkled their houses and stables with holy water as a protective measure.[127] In 1962, the priest at Douarnenez blessed the electric transformers, while today another priest blesses motorcyclists and their powerful bikes. And, with or without clergy, ships are still baptised when they are launched.

Water and standards of health

On a more prosaic level, the improvements in the quality of water that had been made in the course of a century led to a rise in the standard of health of the French people. Thus even before adequate vaccination was extended throughout the country by the devotees of Pasteur, the incidence of typhoid fever had fallen considerably. The typhoid mortality rate in towns with more than 5,000 inhabitants fell rapidly from 5.2 per 10,000 in 1886–90 to 2.4 in 1896. In the army, the reduction was even more spectacular: from 31.2 per 10,000 between 1875 and 1879, it fell to 9.8 in 1895–6.[128] The incidence of diarrhoea and enteritis, which were often very serious illnesses in children under two, fell considerably between 1906 and 1927 from 10 to 3 per 10,000.[129] Similarly, the survey of water purity that was conducted when the local average mortality rate exceeded the national average for three consecutive years (Ministry of the Interior circular of 2 April 1906) contributed to the fight against water of inadequate or suspect quality. Thus, since they had had the simple but brilliant idea of treating the medium in which disease thrived and even before they were assisted by the invention of sera and vaccines, the hygienists had achieved their objectives of improving public health and making France richer and more powerful.

Nevertheless, despite surveys and monitoring, the results obtained did not always match expectations. In 1933, Charles Nicolle warned: 'Typhoid fever is such a widespread disease that it must be considered a constant threat.'[130]

Indeed, half a century after the discovery that typhoid was a waterborne disease, real centres of infection still persisted in France, particularly in Corsica, Languedoc and Provence,[131] but also to a lesser extent in many other areas.[132] However, on the eve of the Second World War, there had been a considerable overall reduction in the incidence of typhoid; public and private hygiene measures (such as disinfection, purification and a reduction in fecal contamination) had been reinforced by regular monitoring carried out by hygiene inspectors (after 1924), by the establishment of health records (after 1909) for local authority districts and individual houses, by health education in schools (after 1883), by press and government propaganda and, above all, by vaccination, which became compulsory in 1914 and generalized by about 1925 and which Charles Nicolle saw as 'a more effective solution, by means of which the consequences of failures, which are still possible with the other methods, can be avoided'.[133]

Thus the advancement in the quality of water and the separation of clean and waste water led to an overall improvement in standards of sanitation in France. From the sanitary and medical point of view, the results were largely positive, which was due largely to modern hygiene and medicine, economic growth, social and sanitary reformism and determinist attitudes to the body and nature.

In keeping with the 'flamboyant' capitalism of the period, an immense task had been achieved, or was at least well on the way to completion. Water had been protected, captured, transported and monitored and was now supplied in all its purity to within easy reach of people's homes, and often right inside them. The removal of cesspits and, more slowly, the building of sewers had remarkable effects on standards of sanitation. Thus the construction of two networks, one for water supply and one for drainage, was accelerated at the beginning of the twentieth century, in accordance with the rule laid down in 1880 by the Academy of Medicine, the wishes expressed in 1884 by the Higher Committee for Public Health in France and the great law of 1902.

The military authorities, more concerned than any government about their population, paid immediate attention to this appeal. Whether they were involved in a colonial war or in a conflict with Germany, the fundamental objective of the authorities, at a time when the infantry was the backbone of the army, was to have at their disposal the greatest possible number of able-bodied men. It was for this reason that the fight to improve sanitary conditions in barracks entered an active phase from 1885 onwards. Monitoring of drinking water was stepped up by means of bi-monthly chemical and bacteriological analyses, which resulted in a considerable fall in the typhoid mortality rate in the army. Nevertheless, in 1896, the typhoid mortality rate in the German army was five times lower than among French troops. According to Professor Brouardel,[134] however, the blame for this cannot be attributed to the French army's sanitary corps, but rather to the fact that water hygiene was generally worse in France than in Germany. Culpability therefore lay with 'civilians', at least if 'army regulations relating to drinking water'[135] had been obeyed by soldiers. . . . At all events, many towns above a certain size experienced a rapid fall in their typhoid mortality rate from the end of the nineteenth century onwards, even though some regions experienced a smaller reduction than others and the countryside lagged behind the towns and cities. It was often the case that:

the elimination of contaminated water sources was in itself sufficient to make a considerable improvement in sanitary conditions for both soldiers and civilians. Thus in 1896 at La Roche-sur-Yon, the well at Erquebouilles was taken out of service; from 1875 to 1886, the typhoid mortality rate in the garrison was 67 per 1,000; between 1887 and 1896, it fell to 4.9. . . In Amiens in 1881, the Marie-Caron fountain was taken out of service and the mortality rate fell from 101 per 1,000 to 6.2 in the army and from 10.8 to 3.5 among the civilian population.[136]

Similarly, for the periods between 1892 and 1894 and 1895 and 1898, the positive impact of water supply projects on the health of the population was evident in about forty local authority districts scattered among seventeen different departments: the overall reduction in mortality was between 5 and 10 per cent, and the number of deaths from typhoid was reduced to virtually zero.[137]

As a result, the life-giving properties of excrement, the protection afforded by dirt, the traditional fear of water and, more especially, of baths, the price demanded by 'purveyors of water' and the complexity of the files that had to be submitted in order to obtain authorization from the Higher Council for Public Health in France and then to apply for state subsidy did not prevent a slow process of change from coming to maturity; nor did it stop the positive health effects of the conquest of water from making themselves felt. The gradual strengthening of state influence over the health of the population, combined with the concern shown by mayors and many citizens with the provision of an adequate supply of pure water and a proper drainage and sewage system for harmful waste, had borne fruit. And Brouardel demonstrated his wisdom when he wrote in 1899 about the threat posed by contaminated water:

Legislation can be introduced, but when it affects the personal habits of an individual's daily life, it can only be effective if public opinion calls for it. Central government has a legitimate right to intervene when the higher interest of the country is at stake, but such intervention can only be effective if public opinion earnestly requests it.[138]

At a late date, often after 1860 and even after 1920, this 'demand' finally had the effects on standards of health in the country that had been anticipated by hygienists.

Mains drainage replaced, albeit slowly, the practice of throwing waste straight into the street. Individual WCs replaced the easily blocked culvert used, among other things, for the disposal of human excrement which had first appeared in Troyes in 1208 and had spread to most towns before the middle of the fourteenth century. The sanitary slop pail, the source of morning conversations about nocturnal excretion, gradually disappeared. Washbasins and baths made an antique, even a museum piece of 'the galvanised tin tub which, at the end of the nineteenth century, represented, even in its most rudimentary form, a modern refinement which still surprised and frightened people somewhat'.[139] Half a century later, however, municipal decrees, such as the one addressed by the mayor of Illkirch-Graffenstaden to the prefect of the Lower Rhine on 6 September 1869, were still not always a dead letter:

We, mayor of the parish of Illkirch-Graffenstaden [. . .]. In the matter of the laws governing the power of the municipal authority and in view of the complaints that have been made to us about pools of stagnant, malodorous water in certain parts of the parish drainage system where there is little natural slope; considering that this state of affairs is due largely to the fact that some householders or tenants allow water from dungheaps or sinks to flow from their yards on to the public highway and that it is in the interests of public hygiene to eliminate these malodorous pools of water on the public highway;
decree as follows:
Article 1. All householders and tenants in this parish are hereby forbidden to drain or to allow to drain from their yard or house on to the public highway any liquid other than rain water or pure well water.
Article 2. Infringements of this order will be noted and pursued in accordance with the law. . . .[140]

A new relationship to water, resulting in a higher standard of sanitation, was thus gradually being established.

The pleasures of water

With all due deference to legislators, hygienists and other well-informed people, water is more than just a bacteriologically pure substance with healing and curative properties. Before it found its way into street fountains and then into taps, it was also a source of pleasures, particularly those of the village fair or even of the traditional great wash.

Even if only for a brief moment, the fair enabled people to forget their conflicts and worries and to lose themselves in its marvels. Heralded by loud peals of bells or the strident brass of the village band, it was a delight for the eye to behold. Water had had a part to play: the streets had been washed down and swept clean. Tapestries, curtains or sheets hung in front of the houses. And in the evening, the illuminated face of a Chinese lantern or a candle would combine water and fire in a visual delight.[141]

'The triumph of water.' Lecture by M. Bichat of the University of Nancy, on the distribution in the city of water from the Moselle, 1879.

(Photograph: © Éditions Robert Laffont.)

In July 1895, at Puisserguier (in the department of Hérault), water was no longer an accessory to the fair but rather the subject of the celebration. According to a contemporary report:

This magnificent fair was presided over by the prefect of the Hérault and the sub-prefect of Béziers; they were welcomed by the mayor, M. Guilhaumon Joseph and his municipal council, the engineers and entrepreneurs who had played a part in the works and the band of the 17th Infantry Regiment.

The organization of this fair had been entrusted to M. Rossi, director of the gasworks. Preceded by the military band, the dignitaries went on a tour of the town, stopping on their way at the public fountains, that had been decorated by the inhabitants of each district. The most beautiful was the Newgate fountain which had been decorated by the local people and Mme Lavigne; it was completely covered by a sheet of red velvet strewn with gold stars. From the steps right up to the Maison Tournal, the Promenade was taken up by a group of wooden huts, decorated with green plants and with red hangings on the walls, in which was served a banquet for more than 200 people [. . .]. After the meal, there was a concert and ball which lasted until 3 o'clock the next morning. And since everything in France ends with songs, Rouquier, our Provençal writer, had composed a song in honour of Puisserguier and its spring.[142]

At the exhibition of 1900, the water tower and the Palace of Electricity were intended to exceed the royal splendours of Versailles and to dazzle the 45 million visitors:

The water tower, made of staff carefully repainted in oils, was built in the style of Louis XV, with very modern and extremely rich ornamentation. Despite its vast bulk (the central recess was thirty-three metres wide and eleven metres deep), its height, its harmonious proportions, the apertures provided on all sides and the perfect streamlining that mirrored so well the movement of water all combined to create an impression of majestic grace [. . .].

The water was fed into the recess in the form of a dolphin's mouth located in the roof. It was supplied from the Villejuif reservoir, from a great height, by means of Worthington pumps positioned on the banks of the Seine and a pump in the basement of the water tower, which conveyed the water from the lower tanks to the upper basin, thus recycling it indefinitely. Despite a consumption of 4,000 cubic metres per hour, the volume of water seemed to be inadequate [. . .].

The water jets were also dwarfed by the vastness of the structure. Although they were higher than those seen in 1889, with some of them reaching a height of twelve metres, they seemed to be somewhat smaller [. . .].

In order to gain an accurate impression of the size of the water pipes, it was necessary to go into the basements underneath the tanks; it was like entering an alien world where the pipes intertwined with each other like enormous snakes [. . .]. The water tower and the tanks also played a part in the illuminations of the Palace of Electricity, indeed they were the most brilliant part of the display. In particular, the vault, studded with incandescent lamps, sparkled like a canopy of precious stones of constantly changing hues.[143]

In this way, the twentieth century revived the traditional festivals of water and light and once again made water into a 'liquid flame'.[144] Whether valued or despised, the pleasures of the senses, and in particular the long-prohibited contact between the body and water, had, after a very long time, been united with the visual pleasure that had been revived by the enchantment of electricity.

It is true that in 1836, 'it was necessary to live in the centre of Paris and be a dandy like Barbey d'Aurevilly in order to pay every morning for a hot bath at the Chinese baths on the Boulevard des Italiens.'[145] Barbey

soaks himself for a long time in order to calm his frayed nerves. If the bath has no effect, he swallows a mixture of his own devising: eau de Cologne mixed with water, or water and vinegar, tea (which he hates, despite his Anglomania), opium, laudanum and kirsch, in order to dull the pain in the small of his back [. . .]. He has himself massaged, perfumed, pomaded, powdered and his hair dyed black as required by fashion and dressed, since dandies wear their hair curled into the shape of a crown.[146]

The bath taken by miners on their return from the pit was less medical, less refined and less complex:

The tub was refilled with warm water and father was beginning to take off his jacket. A warning glance was the signal for Alzire to take Lénore and Henri to play outside. Dad did not like bathing in front of the family, as they did in many other homes in the village. Not that he was criticizing anybody; he only meant that dabbling about together was all right for children.

'What are you doing up there?' Maheude called up the stairs.

'Mending my dress that I tore yesterday,' Catherine called back.

'All right, but don't come down; your father's washing.'

Maheu and his wife were alone. [. . .] He crouched naked in front of the tub and began by dipping his head in, well lathered with soft soap. The use of this soap for generations past had discoloured the hair of all these people and turned it yellow. Then he stepped into the water, soaped his chest, belly, arms and legs and rubbed them hard with his fists. [. . .] She was going on with the usual routine, having rolled her sleeves up so as to do his back and the places that are hard to get at. Anyway, he loved her to soap him and rub him all over fit to break her wrists. She took some soap and worked away at his shoulders, while he stiffened his body to withstand the attack. [. . .] She had gone down from his back to his buttocks and, warming up to the job, she pushed ahead into the cracks and did not leave a single part of his body untouched, making it shine like her three saucepans on spring-cleaning Saturdays. [. . .] She was now drying him, dabbing with a towel at the places that were difficult to dry. He, happy and carefree, laughed out loud and threw his arms around her.

'Don't be silly! You have made me all wet!' [. . .] He seized her again and this time did not leave go. The bath always ended up like this — she made him excited by rubbing so hard and then towelling him everywhere, tickling the hairs on his arms and chest. It was the time when all the chaps in the village took their fun and more children were planted than anyone wanted. At night, they had their families in the way. He pushed her to the table, cracking jokes to celebrate the one good moment a chap can enjoy during the whole day, calling it taking his dessert – and free of charge, what's more![147]

Water and change

It is worth reiterating that 'pure water was for a long time the prerogative of a minority, and until the end of the nineteenth century poor health could be attributed largely to a poor water supply, the high price of that supply and the total absence of hygiene which encouraged the outbreak of epidemics.'[148] It is the duty of the social historian to stress this and to point to the privileges reserved for an elite even well after the 1789 Revolution. Flora Tristan, in her 'Promenades dans Londres', described this in vivid detail:

London does not have any of those lavish monumental fountains that enliven the squares of Paris and speak to everyone in the language of art; in many streets, however, one comes across iron fountains equipped with a pump. An iron chain is attached to the pillar; from the end there hangs a ladle of the same metal. This ladle is the inexpensive cup offered to the poor man by his lord and master the rich man. 'Look, water here costs the people nothing; they can drink it in comfort and without having to fetch it from the river.' This is how well-to-do people in London speak, *yet they never drink water* [. . .]

There was one of these fountains a few yards from my house; all the time I could hear the noise of the chain and the ladle falling back on to the block of stone around the fountain, and I would say to myself: 'That is one of my brothers drinking water, that stale, nauseating London water!' It is true that not all the water supplied to the city comes from the Thames, but all of it is harmful to the stomach and is often the cause of dysentery and fevers! The harsh sound of iron breaks my heart; it resounded in my ears like a funeral knell! Wretched people! Will God leave you to the mercy of your lords, who, devoid of pity, watch you die that slow, agonising death which, every hour and every minute, kills the victim who struggles vainly in his agony?

However, the reports of the committees of hygiene and of the factory inspectorate, the files of technical departments and papers given at conferences and to learned societies all bear witness to the birth of a veritable 'cultural revolution', a revolution that the nation's elite had for a long time appropriated for its own gain. As it emerged from streams, water brought with it rumours of salvation.

Thus the significance of the relationship with water tended to be reversed as an ethnographic culture gave way to a scientific culture: water was no longer to be the medium for the journey into the hereafter, but rather a manifestation of vigour, fertility and health. The distinction between healthy and unhealthy, which had existed in outline at the end of the *ancien régime*, finished by transforming the anthropological landscape of France with respect to beliefs, knowledge, rites and objects.

Science had identified the boundary between pure and impure water; as a consequence, water was now controlled and entrusted with the task of carrying away waste and excrement through underground pipes; dunghills were no longer placed in close proximity to wells, cesspits became leakproof, if not septic, and just as in earlier times cemeteries had been taken outside the parish boundary wall.

Those whose role it was to manage and manipulate space – architects, town planners, sanitary engineers, hygienists, chemists and engineers – created new objects and sculpted new structures, concealed the hydraulic systems and took water underground in order to protect it and to protect man from it; then, when it had performed – or failed to perform – its cleansing function, they sent it back to the river or made it gush forth from fountains or even from taps and the new British-style lavatories. In their role as conquerors, they planned, organized and built a new 'body', that of water, which they incorporated into nature in the image of man.

Thus, by a process of descent down the urban hierarchy and of social capillarity, and as a result of the rural exodus and the repeated efforts of schools, the medical profession and the media, the old relationship with water was transformed so that it could exert an influence in accordance with the new sanitary code, which had become the only operative criterion. Assisted by administrative centralization and increasing state intervention in health matters, the policy of pious wishes that prevailed between 1750–1880 gave way to the construction between 1880 and 1940 of the socio-sanitary sphere.

The power of water, now that it had once again been captured and tamed, charged for and made accessible to an ever-increasing number of people, was further strengthened by knowledge, exhibitions, monumental fountains, advertising and education. Moreover, water was now more diffused than it had ever had been. This reflects the importance of this element in any reading of the history of a culture and society, even before it became the subject of religious and social sacralization.

Tamed by the successors and rivals of the water diviners, faith healers, priests and the 'old wives', water is still symbolic of the transition to cleanliness and hygiene and has been domesticated to the point of becoming a daily habit. Even though it has been secularized and made commonplace, this 'product' of the industrial revolution, because of its central place in nature and its relationship with our bodies, has retained its sacred character as it has infiltrated our customs, insinuated its way into our houses, bathed our inner depths and given rise to new rituals.

Because washing, particularly after 1940, only rarely symbolized the continuity between life and death, but rather a break based on the distinction between the healthy and the unhealthy, the language of water sought to exclude death which, at precisely this period, was becoming privatized and a taboo subject and which only the conquest of water succeeded to a certain extent in taming. The long-established success of toilet waters (eau de Cologne was originally a remedy against the plague), the fashion for bathing in the sea, the popularity of hydrotherapy and the current vogue for winter sports cannot be explained in any other way: water is still a resource which a culture which is more polysemous than scholarly or popular assimilates in 'layers', rather as one digests the various courses of a meal.[149]

CONCLUSION

The Significance of Change

Knowledge of the microbial aetiology of typhoid fever and the way in which it is spread forced the political and scientific authorities into sudden awareness of the problems of sanitation and public health. On their joint command, an entire sanitary system based on the newly discovered hygiene standards was put in place: water was now monitored, distributed and drained away. As a result, the environment underwent a profound change, particularly in towns and cities, which now had a supply of pure water and a network of sewers for the disposal of liquid and solid waste.

The design and construction of dwellings also underwent change in order to accommodate the new water supply and drainage systems, and craftsmen and engineers created a whole host of new products. However, it was not only the material world that was changed. As hygiene standards were internalized, a new code of behaviour and new sets of attitudes emerged. Both the body and the physical environment had undergone a radical transformation.

In the first instance, the notion of water (and of purity) had evolved. There was no longer any question, as there had been in earlier times, of trusting appearances: running water no longer symbolized life, and stagnant water no longer represented death. It was now clearly understood that even running water might contain germs that could cause disease and even death. It seemed that microbiology had consigned the traditional symbolism of water, and with it a long-established, essentially agrarian culture, to the antiquities department.

Moreover, it was now recognized that the nature of water could vary infinitely without these fluctuations being ordinarily perceptible. These variations could be dependent on location, which might mean that water previously deemed pure because it came from a spring would be considered impure, and also on surroundings, and in particular on proximity to bodily waste (dungheaps and excrement). This new realization marked a change from the popular beliefs of earlier times, which had stressed the life-giving power of such filth of which peasant farmers were aware from their experience on the land.

A third factor which contributed to the change in the perceived nature of water was the passage of time, which, on the one hand, strengthened the deep-rooted belief in the rhythm of the seasons, while on the other stretching to breaking point the traditional relationship between Jack Frost and renewal; water and the microbes it contained could strike at any moment, unrestricted by any cosmic relationship. And yet attempts had to be made to escape from it.

The knowledge of those doctors who followed in Pasteur's footsteps thus marked a break with the past. It was a decisive factor in the acceleration of a process which had been under way for a long time. Not only did it bring about a change in the way that water was taken over by technology and the economy, together with an even more lasting transformation of the structure of dwellings and of the great centres of population; it also provided the intellectual framework and established the cultural norms which were to have a lasting influence on perceptions of the relationships between the body and nature.

The revolution in our knowledge of water and the discovery that it was the source of numerous diseases had social effects of considerable importance. Cleanliness became confused with hygiene, if not with purity, to the extent that the medical profession made haste to declare that the water of Lourdes had no therapeutic value.[1] The profession declared itself to be the sole custodian of truth and laid down standards and patterns of behaviour which led to a proliferation of devices and rituals intended to banish dirt, disease and death by declaring war on the enemy within – microbes – at a time when attempts were being made to improve the health of the people in order to fight the enemy without, namely Germany.

This medical knowledge acted upon the world in the same way as a surgeon operates on a body. It too required instruments. For this reason, it was transmitted by schools, official and otherwise, by the middle-class and popular press and by the trade union movement (particularly the reformist unions) which supported hygiene as a means of social and medical advancement. In primary schools, where education had been free of charge, secular and compulsory since the time of Jules Ferry (1881), lessons in the catechism were replaced by lessons on hygiene. The popular press read by craftsmen, manual workers and peasants, which came into being in 1863 with *Le Petit Journal*, diffused the new model, overlaid with the cheap finery of bourgeois morality.

Typhoid, 'the disease of dirty hands', was included in the war waged on microbes and on the dust which brought them into being. The 'cleanliness checks' carried out each morning by primary school teachers and the 'presentation of hands' (in itself a religious gesture), which was intended to punish those with dirty nails, were included in the new school rituals. The equipment, articles and water intended for washing and which had for a time become anti-contagious and antiseptic, even recommended by Pasteur, came into use in the home, replacing the domestic filters which had earlier been the prerogative of the elite classes. Washing one's hands before sitting down to a meal was deemed hygienic and stipulated for that reason. For example, the ancient custom of washing the hands on returning from a funeral still persists today. The factory was seen as place where peasant farmers recently arrived from the country could receive health education; workers' dwellings, initially in the form of collective housing and then of the individual house, were promoted by government and employers, who were anxious to maintain the good order, cleanliness and health of the work force, which had to be kept docile, healthy and productive.

Seized upon by the dominant classes and diffused among the dominated masses, Pasteur's discoveries belonged nevertheless to a much older tradition. Undoubtedly, a sudden change in the conception of water transformed the scenery, not only in the home and the workplace, but also with respect to body maintenance. Objectified by scientific knowledge, water was secularized and transformed into an everyday consumer product in a rich and temperate country which called itself secular. Nevertheless, it retained its original ambiguity. It remained a 'substance of life and death', in the words of Gaston Bachelard, the subject and object of a long dialogue in which culture, desires and anxieties were apparent and which reflected the intimate relationships between human beings and the world (the desire to defend oneself against death by banking on fertility is one aspect of this). The aetiology may have changed and the criteria by which the purity of water was judged may have been transformed, but water remained ephemeral and eternal, a symbol of the transience of life.

The gradual transition from an ethnographic to a scientific culture was not without its casualties: links were slackened and the sociability associated with some of the ritual uses of water disappeared. And sometimes partial or even total amnesia was induced: the link that water provided between the body and nature, between creation and the creator, between the generations, between the internal and external worlds and between the visible and the invisible was in danger of disappearing.

Water had become a sanitary, industrial and commercial product, a raw material, a source of energy and a target for advertising. The actions associated with the use of water were secularized, almost vulgarized, made commonplace and mechanized, to the point where some people forgot entirely that they were participating in an eternal ritual intended to banish the spectres of death, to slay the phantasms of the night and to calm sexual passion.

Because the measure of time had changed and because productivism had become the *ne plus ultra* of a society in the clutches of 'industrial fever', some people were able to believe and to lead others to believe that water was no longer ambivalent, that it was only a divisible, measurable and profitable element of the capitalist economy, whereas it was quite evident that it provided a considerable amount of information on the society and culture of nineteenth-century France.

Undoubtedly, the actions associated with water, those linked to work and washing among others, no longer seemed to be in direct contact with cosmic and calendar time, nor with the roles traditionally ascribed in our society to each social group and each of the two sexes. Thus the 'great spring clean', the collective washing of clothes, either annual or seasonal, the parsimonious use of water and of baths, the daily quest for water which made Victor Hugo and his readers weep for the fate of little Cosette, in short, the social life of the village that centred around the well and the fountain, to say nothing of the rituals of healing and witchcraft associated with them, all belonged to a past age that some people, attached to their distant youth, saw as a 'golden age'. And because the ethnographic culture possessed 'the beauty of death', virtually nobody today escapes its charm.

However, such an attitude reflects a tendency to glorify the past, if not to fall prey to an unhealthy attachment to it. Moreover, this attitude may well prevent us from turning to account the particular attention that anthropology brings to bear on the relationships between the body and nature in an attempt to interpret our recent and present culture. Indeed, in an old country like France, impregnated with Christianity and endowed with a long memory, whatever some may say in this respect, the representation of water remains characterized by the notion of contamination and pollution which concerns the body as much as nature, to the extent that water is an element in the union between the two.

In fact, examination of advertising from the recent past, particularly that relating to mineral, toilet and healing water, of etiquette lessons, of handbooks of military instruction and of school hygiene rules and moral precepts reveals the existence of three cultural 'strata', the common aim of which was 'healing' and survival.[3]

To describe the situation in nineteenth-century terms, the most recent of these strata, the so-called 'rationalist stratum', declared itself, in its scientific blindness, to be the only champion of reason. The two others, the 'symbolic' and the 'religious' strata, which were closely related to each other, were relegated to a very lowly status, if not that of ancient superstition and contemptible obscurantism.

In fact, however, all three overlapped closely. The most recently established, which sought to turn hygiene into a science and failed of course to achieve this aim since it was in fact a cultural code, made its mark as much through the power of the symbols and issues for which it was the medium as through the scientific notions which it espoused.

In reality, it was through new 'prejudices', new knowledge, new practices, new rituals (those associated with hygiene), new aims which served to justify instances of designation, transgression and exclusion and through new prayers and incantations contained in the messages propagated by advertising, schools and health officials, that French people in the contemporary era appealed to the power with which they invested themselves and which they attributed to water in order better to conquer it, to allay their anxieties and to bring forth their pleasure.

Natural, supernatural, artificial, developed and not created, water remained changeable yet eternal. In the nineteenth century, in the age of scientific triumphalism, its power, particularly its 'healing' power, was all the greater since Pasteur and Koch had demonstrated and, in certain instances, eliminated its harmfulness.

With the assistance of 'flamboyant' capitalism, which was all too happy to find in it a new source of profit, it remained, even in the middle of the nineteenth century, the subject of secrets[4] and conflict, on the social, political[5] and cultural[6] levels.

Subjugated, domesticated, mechanized and made profitable, it imposed itself on a rapidly expanding population. It thus revealed a different knowledge, increased power and a new 'image of the world'. In its own way, it helped to define 'man's real nature, which is not so much to make tools or to be reasonable as to create himself and to associate with others, in short to develop his natural state'.[7]

Notes

General Introduction

1 According to a survey carried out by I.P.S.O.S. in 1983, reported in *Marie-France* (February 1984), 38 per cent of women and 22 per cent of men in France wash all over every day.

2 Claude Duneton, *Je suis comme une truie qui doute* (Paris, 1976), p. 74.

3 Victor Hugo, *Les Misérables* (London, 1976), p. 351.

4 J.-F. Blondel, *Traité de la décoration* (Paris, 1776), vol. 2, p. 402.

5 N.-E. Restif de la Bretonne, *La Paysanne pervertie* [1784] (Paris, 1972), pp. 422–3.

6 Guy Thuillier, *Pour une histoire du quotidien au XIXè siècle en Nivernais* (Paris-La Haye, 1977), p. 20.

7 Louis-Sébastien Mercier, *Tableau de Paris*, ch. 671, vol. 2, pp. 339–41.

8 Thuillier, *Pour une histoire du quotidien*, p. 15.

9 Gaston Cadoux, *La Vie des grandes capitales* (Paris-Nancy, 1913), 2nd edn, pp. 114–15.

10 Françoise Loux, 'Transmission culturelle chez les catholiques et les protestants: les soins corporels à Chardonneret', *Ethnologie française*, 1–2 (1974), pp. 145–8.

Part I

1 This theory of the three 'ages of mankind', already put forward by Auguste Comte, was taken up again a few years ago by Jacques Gélis and expounded at my seminar at the Ecole des Hautes Etudes en Sciences Sociales on the history of the body and childbirth.

2 Quoted in Wanda Bannour, *Eugénie de Guérin ou une chasteté ardente* (Paris, 1983), p. 107.

3 *Le Médecin des dames, ou l'art de les conserver en santé*, {1772} (Paris), new revised and expanded edition p. 317.

4 Philippe Perrot, *Le travail des apparences. Ou les transformations du corps féminin XVIIIè–XIXè siècles* (Paris, 1984), pp. 13ff.

Chapter 1 New knowledge

1 Cf. *Histoire critique des pratiques superstitieuses* (Rouen, 1702), pp. 36–7.

2 Quoted by P. Chassé and N. Watrin in their unpublished final-year dissertation at the Ecole nationale des Travaux Publics de l'Etat, 'Le sourcier, homme de passé ou d'avenir' (Vaux-en-Velin, 1982).

3 Cf. Yves Rocard, 'Le signal du sourcier', *La Recherche*, 124 (July-August 1981).

4 André Guillerme, *Les temps de l'eau. La cité, l'eau et les techniques* (Seyssel, 1983), pp. 190ff.

5 André Guillerme, *Réseaux hydrauliques urbains*, doctoral thesis, Université de Paris VIII (Paris, 1981), vol. 1, p. 390.

6 S. Parkes, *The Chemical Catechism* (London, 1818) 8th edn, p. 4.

7 A. Dechambre et al., *Dictionnaire encyclopédique des sciences médicales* (Paris, 1885), p. 352.

8 Charles Lucas, *Essay on Waters* (London, 1756), vol. 1, p. 81.

9 Ibid., p. 127.

10 Joseph Browne, 'Memoir of the utility and means of furnishing the city with water from the River Bronx' (1798) in *Report of the Committee on Fire and Water relative to introducing water into the City of New York*, Documents of the Board of Alderman and Board of Assistants, no. 61 (1831), p. 266.

11 Claude Grimmer, *Vivre à Aurillac au XVIIIè siècle* (Aurillac, 1983), p. 100.

12 Archives of the Royal Society of Medicine, document 18, Rennes, 25 November 1789.

13 Ibid., box 177, file 1, item 5, Valence, 1 September 1789.

14 Ibid., box 177, file 1, item 5, Thiers, 4 February 1788.

15 Lepecq de la Clôture, *Collection d'observations sur les maladies et constitutions épidémiques* (Rouen, 1777), p. 259.

16 Roger Chartier, 'L'Académie de Lyon au XVIIIè siècle', in *Nouvelles études lyonnaises* (Genève, 1969), pp. 249–50.

17 National Archives, F^{12} 2408, letter from Sieur Garroy to the Consultative Committee on Arts and Manufactures, 3 May 1815.

18 National Archives, F^{12} 2408, Paris, 16 March 1819.

19 F. Diénert, *Eaux douces et eaux minérales* (Paris-Liège, 1912), p. 5.

20 Archives of the City of Brussels, 2 November 1860, Report of the medical committee on the analysis requested by the City of Brussels on 26 October 1860, quoted in Liliane Viré, *La Distribution publique d'eau à Bruxelles 1830–1870* (Brussels, 1973), p. 36.

21 Ibid., p. 36.

22 Teissier Roland, (1844), vol. 3, p. 717.

23 S.E. Finer, *The Life and Times of Sir Edwin Chadwick* (London, 1952), p. 230–43.

24 Nelson Manfred Blake, *Water for the cities. A History of the Urban Water Supply in the United States* (Syracuse, 1956), p. 46.

25 Heidrun Winkler, *Wasserversorgung und Abwasserbeseitigung als Probleme der Bielefelder Stadtpolitik in der zweiten Hälfte des 19. Jahrhunderts* (Bielefeld, 1986), pp. 26ff.

26 J. Engelmann, 'Geschichte der Gewerbetätigkeit Kölns', in *Köln und seine*

Bauten (Cologne, 1888), pp. 725–53. Work quoted by Klaus Norbisath, 'Geschichte der Wasserversorgung der Stadt Köln', inaugural dissertation, Faculty of Medicine (Cologne, 1970), pp. 69–72.

27 Bontron and Boudet, *Hydrométrie* (Paris, 1856), p. 10.

28 Seeligman, *Essai chimique sur les eaux potables*, 1st. report, (Lyons, 1860), pp. 40ff.

29 Peligot, 'Recherches sur les matières organiques contenues dans les eaux', *Annales de chimie et de physique*, Oct. 1864, p. 225. Twenty years later the upper limit recommended by hygienists was 21°.

30 Houzeau and Blavier, *Rapport au maire de la ville d'Angers* (Angers, 1853), p. 41.

31 Viré, *La Distribution publique d'eau à Bruxelles 1830–1870*, p. 189.

32 Ibid., p. 190.

33 Ange-Pierre Leca, *Et le choléra s'abbatit sur Paris, 1832* (Paris, 1982), p. 200.

34 Ibid., p. 262.

35 Ibid., p. 264.

36 *Report on the chemical quality of the Supply of Water to the Metropolis, B.P.P.* (1851), XXIII, pp. 9–10.

37 Ibid., pp. 8–9.

38 Norbisrath, 'Geschichte der Wasserversorgung der Stadt Köln', pp. 74–8.

39 Letty Anderson, 'Incendie et maladie. Le développement des réseaux hydrauliques en Nouvelle-Angleterre 1870–1900', *Les Annales de la Recherche Urbaine*, 23–4, (July–December, 1984), p. 56.

40 Charles E. Rosenberg, *The Cholera Years. The United States in 1832, 1849 and 1866* (Chicago and London, 1962), p. 121.

41 Ibid., p. 127.

42 Ibid., pp. 121–2.

43 Ch. Achard (Dr), *Les maladies typhoïdes. Etudes cliniques, pathologiques et thérapeutiques* (Paris, 1929), p. 27.

44 Robert Garry (Dr), 'Contribution à l'histoire de la fièvre typhoïde', medical thesis (Rennes, 1960).

45 P. Brouardel and L.H. Thoinot (Drs), *La fièvre typhoïde* (Paris, 1895), pp. 12ff.

46 Ibid., p. 33.

47 Cf. Alice Peeters, 'L'hygiène et les traditions de propreté. L'exemple des Antilles françaises', *Bulletin d'ethnomédecine*, 11 (March 1982).

48 André Guillerme, 'Capter, clarifier, transporter l'eau – France 1800–1850', *Les Annales de la recherche urbaine*, no. 23–4 (July–December 1984), pp. 31–46.

49 Thuillier, *Pour une histoire du quotidien*, 'L'eau', pp. 11–30.

50 Genieys, *Essai sur les moyens de conduire, d'élever et de distribuer les eaux* (Paris, 1829), p. 155.

51 Darcy, *Recherches expérimentales relatives au mouvement de l'eau* (Paris, 1857), vol. 1, ch. 6.

52 Doyen, *Histoire de Beauvais* (Beauvais, 1843), vol. 2, pp. 182–5; quoted in Guillerme, 'Capter, clarifier, transporte l'eau – France 1800–1850'.

53 According to Héricard de Thury, quoted in Guillerme, 'Capter, clarifier,

transporter l'eau – France 1800–1850.

54 Darcy, *Les fontaines publiques de la ville de Dijon* (Paris, 1856), p. 245.

55 Gaudin, 'Note sur une conduite d'eau de la ville et du port de Cherbourg', *Annales des Ponts-et-Chaussées* (1836), 2, p. 352.

56 Period 1884–1902; sample of 25 departments and 302 schemes; National Archives, F^8 179–206.

57 Alain Corbin, *Le Miasme et le Jonquille* (Paris, 1982), p. 249.

58 Ibid., p. 248.

59 A. Calmette (Dr), *Recherches sur l'épuration biologique et chimique des eaux d'égout* (Paris, 1905).

60 Alfred Franklin, *La Vie privée d'autrefois. Moeurs, modes, usages des Parisiens du XIIè au XVIIIè siècle* (Paris, 1890).

61 Pierre Boutin, 'Points de repère pour une histoire de l'assainissement', CEMAGREF, BI, 314–15 (March–April 1984), p. 43.

62 *Royal Commission on river pollution* (BPP, 1874), XXXIII, p. 471.

63 Viviane Claude, 'Strasbourg 1850–1914, Assainissement et Politiques urbaines', thesis, Ecole des Hautes Etudes en Sciences Sociales (Paris, 1985), 2 vols, vol. 1, pp. 83–6.

64 Letter from M. von Pettekoffer to the Alsatian hygienists, *Archiv für öffentliche Gesundheitspflege in Elsass-Lothringen* (1884), p. 94, letter quoted in Ibid., vol. 2, p. 447, n. 54.

65 René Dubos (Dr), *Mirage de la santé* (Paris, 1961), p. 123.

66 Nivet (Dr), *Rapport sur l'engrais humain, les égoûts et les fosses d'aisances* (Paris, 1882).

67 A. Ronna, *De l'utilisation des eaux d'égoûts en Angleterre* (Paris et Liège, 1866).

68 Pierre Boutin, 'Eléments pour une histoire des procédés de traitement des eaux résiduaires', *La Tribune Cebedeau*, 39, 511/512, pp. 3–4.

69 Ed. Imbeaux (Dr), 'Les avantages et les inconvénients des égoûts du système unitaire et du système séparatif', International Conference on Hygiene, (Brussels, 1903), 3rd section pp. 12–13.

70 In 1896, of 708 towns with more than 5,000 inhabitants, 465 had a sewage system and of the remaining two-thirds (465), 301 had two separate networks (the great majority) or a combined system (United States).

71 Dr Imbeaux et al., *Annuaire (. . .) des distributions d'eau de France, Algérie et Tunisie, Belgique et Suisse* (Paris, 1903).

72 Gabriel Dupuy and Georges Knaebel, *Assainir la ville hier et aujourd'hui* (Paris, 1982), p. 4.

73 Irène Lentin, *Attitudes face au péril, à Paris du XIVè siècle à l'aube du XIXè siècle* D.E.A., E.H.E.S.S., (Paris, 1977), p. 5.

74 'Barathrum': the Latin form of a Greek word which referred in particular to a ravine in Athens where the bodies of people condemned to death were thrown.

75 Hugo, *Les Misérables*, p. 1068.

76 Extract from an article that appeared in 1882 in *XIXè siècle*, quoted in Paul Wéry, *Assainissement des villes* (Paris, 1898), pp. 28–31.

77 Gérard Jacquemet, 'Urbanisme parisien: la bataille du tout-à-l'égout à la fin du XIXè siècle', *Revue d'Histoire moderne et contemporaine*, 26 (October–

December 1979), p. 505. Cf. also Roger-Henri Guerrand, 'La bataille du tout-à-l'égout', *L'Histoire*, 53 (February 1983).

78 E. Vallin (Dr), 'L'utilisation agricole des eaux d'égout de Paris et l'assainissement de la Seine', *Revue d'hygiène et de police sanitaire*, 1888, p. 97.

79 P. Koch, 'Etude sur le calcul des ouvrages d'évacuation en fonction de ruissellement dans l'assainissement urbain', *Annales des Ponts-et-Chaussées*, IV (1930), pp. 5–41.

80 *Mémoire sur les eaux de Paris, présenté à la Commission municipale par Monsieur le Préfet de la Seine*, 4 August 1854, Paris, p. 53.

Chapter 2 New objectives

1 M. Eleb-Vidal, *In extenso*, 2 (1984), p. 66.

2 Ph. Ariès, *L'enfant et la vie familiale sous l'Ancien Régime* (Paris, 1960), p. 248.

3 Norbert Elias, *La Civilisation des moeurs* (Paris, 1973), p. 241.

4 Ibid., p. 242.

5 Gilbert Gardes, 'L'eau, de la technique à la poétique', *Monuments historiques*, 122 (August–September 1982), p. 7.

6 Ibid., p. 7.

7 Christian Lorgues, 'L'alimentation en eau des villages de Provence', *Monuments historiques*, 122 (August–September 1982), pp. 33–35.

8 Yvonne Verdier, *Façons de dire, façons de faire. La laveuse, la couturière, la cuisinière* (Paris, 1979), p. 121.

9 Cf. Denis Grisel, 'Les fontaines – lavoirs en Haute-Saône au XIXè siècle', *Monuments historiques*, 122 (August–September, 1982), pp. 11–20.

10 Michelle Perrot, 'Femmes au lavoir', *Sorcières*, 19, p. 128.

11 Arlette Farge, *Vivre dans la rue à Paris au XVIIIè siècle* (Paris, 1979), pp. 48, 50 and 51.

12 Ibid., p. 53.

13 Flora Tristan, *Le Tour de France* (. . .) (Paris, 1973), p. 216. Cf. also P.J. Hélias, *Le Cheval d'orgeuil* (Paris, 1975), pp. 239–40.

14 Françoise Le Brenn, 'Comment blanchir le peuple. Les bains-lavoirs du Temple', *Monuments historiques*, August–September 1982, pp. 30–2.

15 Perrot, 'Femmes au lavoir', p. 129.

16 Ibid., p. 129.

17 Thuillier, *Pour une histoire du quotidien*, 'La lessive', p. 133, n. 44.

18 Ibid.

19 Daniel Halévy, *Visite aux paysans du Centre* (Paris, 1975), p. 161.

20 In these iron boilers, it was the action of the steam that washed the clothes.

21 Verdier, *Façons de dire, façons de faire*, p. 110.

22 Cf. André Burgière, *Bretons de Plovézet* (Paris, 1975), p. 161.

23 Verdier, *Façons de dire, façons de faire*, p. 115.

24 Cf. Maris-Cécile Riffault, 'De Chaptal à la mère Denis: histoire de l'entretien du linge domestique', *Culture technique*, 3 (15 September 1980), pp. 256–63.

25 Cf. Leonardo Benevolo, *Aux sources de l'urbanisme moderne* (Horizons de France, 1972), pp. 91ff, plans pp. 96–7; and cf. Jean-Baptiste Godin (1819–1888), *La Richesse au service du peuple: le familistère de Guise* (Paris, 1874).

26 Perrot, 'Femmes au lavoir', p. 131.

27 Emile Zola, *L'Assommoir* (London, 1958), pp. 137–8.

28 Cf. Damièle Voldman, 'Eaux de Mars', *Monuments historiques*, 122 (August–September 1982), pp. 42–5.

29 Until 1623, the year that the aqueduct was brought into service, the whole of the Left Bank depended for its water supply on water from the Seine and from contaminated wells.

30 Cf. Jean-Louis Hannebert, 'L'aqueduc de Marie de Médicis à Paris', *Monuments historiques*, 122 (August–September, 1982), pp. 17–19.

31 Marc Gaillard, 'Des châteaux d'eau', *L'Express*, 1013 (7–13 December 1970).

32 Georges Duval, 'Les châteaux d'eau', *Monuments historiques*, 122 (August–September, 1982), pp. 37–8.

33 Dominique Jarrassé, 'La ville et les jeux d'eau au XIXè siècle', ibid., pp. 74–7.

34 Maurice Agulhon, 'Les fontaines de village dans la tradition provençale', *Provence historique*, 93–4 (1974); and Agulhon, 'Imagerie civique et décor urbain', *Ethnologie française*, 5, (1975), pp. 33–57.

35 Cf. Daniel Roche, *Le Peuple de Paris. Essai sur la culture populaire au XVIIIè siècle* (Paris, 1981), p. 157.

36 Cf. Corbin, *Le Miasme et le Jonquille*, p. 86; cf. also Perrot, *Le Travail des apparences*, p. 14.

37 Verdier, *Façons de dire, façons de faire*, p. 121.

38 Ibid., p. 121, n. 1.

39 Roche, *Le Peuple de Paris*, p. 159.

40 Ibid.

41 Ibid.

42 According to Catherine Virole, 'Médecins et chirurgiens à Paris dans la seconde moitié du XVIIIè siècle', master's thesis (Paris I, 1977), p. 105.

43 According to Julia Csergo, *Liberté, Egalité, Propreté. La morale de l'hygiene* (Paris, 1980), pp. 329–30.

44 Memoirs of the Comtesse de Pange, née Pauline de Broglie, quoted by Paul Leuillot in his preface to Thuillier, *Pour une histoire du quotidien*, pp. XX–XXI.

45 Georges Vigarello, *Le Propre et le Sale* (. . .) (Paris, 1985), p. 204.

46 Ambroise Tardieu, *Dictionnaire d'hygiène publique et de salubrité* (Paris, 1862), 2nd edn, vol. 2, pp. 184–5

47 Cf. 'L'évolution des conditions de logement en France depuis cent ans', *Etudes et conjoncture*, October–November 1957.

48 *Annuaire statistique de l'INSEE*, 1972, p. 102 (Statistical Yearbook published by the French Government Statistical Service).

49 According to the *Grand Larousse du XIXè siècle*, it is Italian in origin. According to Littré, its etymology is Celtic and is derived from *bideach*, meaning very small. Some English, Italian and Spanish dictionaries note

that the French word has been retained in these three languages: *bidè* in Italian, *bidet* in English and *bidé* in Spanish.

50 Quotation taken from Roger-Henri Guerrand, 'L'âge d'or du bidet', *L'histoire*, 57 (June 1983), pp. 84–7.

51 D'Argenson, *Mémoires*, vol. 1, p. 205. After Henry Havard, *Dictionnaire de l'ameublement et de la décoration. Depuis le XIIIè siècle jusqu'à nos jours* (Paris, 1887), vol. 2, col. 954. On the history of the bidet and the bathroom, cf. Lion Murard, Patrick Zylberman, 'Buanderies de la chair:' *Archithese*, 1 (January–February 1985), 'Sauberkeit/Hygiène' [sic], pp. 25–31.

52 Franklin, *La Vie privée d'autrefois*, 'L'hygiène', p. 32.

53 Ibid.

54 A.J.B. Parent-Duchâtelet, *De la prostitution dans la ville de Paris* (Paris, 1836), vol. 1, p. 135.

55 G. Fustier, *Supplément au dictionnaire de la langue verte d'A. Delvau* (1887), p. 507.

56 Paris, 1912, vol. 10, ch. 13.

57 *Comment installer sa maison* (Paris, 1918).

58 Arthur Young, *Voyages en France 1787, 1788, 1789*, vol. 1, *Journal de Voyages* (Paris, 1976), p. 485.

59 Thesis in progress on bedrooms in nineteenth-century Paris by Rivka Bercovici (sample of 1,000 Parisians representative of the middle classes).

60 E. and J. de Goncourt, *Journal* (1895), p. 854.

61 Comtesse de Pange, née Pauline de Broglie, *Comment j'ai vu 1900* (Paris, 1962), vol. 1, p. 196.

62 In 1962, only 30 per cent of principal residences in France had a washbasin, bath and bidet.

63 Peter Reinhart Gleichmann, 'Des villes propres et sans odeur. La vidange du corps humain: équipements et domestication', *Urbi*, V (1982), pp. LXXXVIII–C. For more detailed studies, see Roy Palmer, *The Water Closet, A New History* (Newton Abbot, 1973); and Roger-Henri Guerrand, *Les lieux. Histoire des commodités* (Paris, 1985).

64 Franklin, *La Vie privée d'autre fois*, 'L'Hygiène', p. 31.

65 Ibid., p. 19.

66 See, in particular, the decrees of 8 November 1522 and 26 August 1531, as well as the one issued in 1780, which made any throwing of 'water, urine, faeces and other waste' punishable by a fine of 300 francs.

67 A.J.B. Parent-Duchâtelet, *Essai sur les cloaques ou égouts de la ville de Paris* (Paris, 1824), p. 211, n. 7.

68 Raymond Collier, *La Vie en Haute-Provence de 1600 à 1850* (Digne, 1973), p. 218, (Archives of the Department of the Alpes de Haute-Provence, 1 B 1036).

69 Molière, *L'Etourdi*, act III, scene XIII (1653)

70 Scarron, *Don Japhet D'Arménie*, act IV, scene VI, (1653).

71 The term 'privy' denotes a private place and, more particularly, a toilet.

72 Saint-Simon, *Mémoires*, vol. 4, p. 386 (1705–9).

73 Saint-Simon, *Lettres*, 20 September 1714, vol. 1, p. 145.

74 Saint-Simon, *Lettres*, 11 June 1720, vol. 2, p. 242.

75 Saint-Simon, *Confessions*, Book 1.

76 Saint-Simon, *Oeuvres*, vol. 13, p. 254.

77 Voltaire, *Oeuvres*, Garnier, vol. 35, p. 73.

78 Young, *Voyages en France*, vol. 1, p. 485.

79 Mercier, *Tableau de Paris*, vol. XI, p. 54.

80 Labarraque, Chevallier, Parent-Duchatelet, 'Rapport sur les améliorations
    à introduire dans les fosses d'aisances, leur mode de vidange, et les voiries
    de la Ville de Paris', taken from *Annales d'hygiène publique et de médecine
    légale*, vol. 14, part 2, p. 5.

81 Ibid., p. 6.

82 Racinais, *Un Versailles inconnu. Les petits apartements des rois Louis XV et
    Louis XVI au château de Versailles* (Paris 1950). See also Félix Gaiffe,
    L'envers du grand siècle (Paris, 1924).

83 Elias, La Civilisation des moeurs, pp. 185–204.

84 *Rapport sur la salubrité des constructions* (Paris, 1881), pp. 10–11.

85 P. Brouardel (Dr), *Rapport sur l'assainissement de Paris* (Paris 1881), pp.
    31ff.

86 Piorry, *Dissertation sur les habitations privées* (Paris, 1837).

87 *Dictionnaire des Arts et Manufactures* (Paris, 1853-5), 'Insalubres'.

88 E. Vallin, *Traité des désinfectants* (Paris, 1883), p. 642.

89 Pan-closets were single-valve and valve and siphon water closets.

90 By about 1880, approximately 50,000 of the 80,000 houses in Paris were
    connected to the city water supply. Léon Colin, *Paris. Sa topographie, son
    hygiène, ses maladies* (Paris, 1885), p. 144.

91 *Sanitary Record*, 52, (1883), p. 188.

92 Latrines, cesspits, soil tubs, barrels, so-called 'Turkish loos', or seatless
    lavatories, close-stools and (more or less) hygienic buckets. Cf. H. Napias
    and A.J. Martin, *L'Etude et les Progrès de l'hygiène en France de 1878 à 1882*
    (Paris, 1882), pp. 183–209.

93 François-Yves Besnard, *Souvenirs d'un nonagénaire* (Paris, 1880), vol. 1, p.
    210.

94 A. Dechambre et al., Dictionnaire de Médecine (. . .), 1885, article
    entitled 'Eau', p. 559.

95 J.-B. Fonssagrives (Dr), *Dictionnaire de la Santé* (. . .) (Paris, 1876), p.
    383.

96 Ibid., p. 382.

97 Dechambre et al., pp. 560–1.

98 Fonssagrives, *Dictionnaire de la Santé*, pp. 382–3.

99 Cf. 'Tableau dressé par Henri Sauvage, architecte', *L'Illustration*, 4491 (30
    March 1929).

100 Washing equipment was very rarely listed in post-obit inventories in Paris
    in about 1780. Daniel Roche, *Le Peuple de Paris*, p. 163, n. 62.

101 Riffault, 'De Chaptal à la mère Denis'.

102 Pierre Pierrard, *La vie ouvrière à Lille sous le Seond Empire* (Paris, 1965), p.
    82.

103 Quotation from Mercier, *Tableau de Paris* (about 1789), taken from A.
    Franklin, *La civilité, l'étiquette, la mode, le bon ton du XIIè siècle au XIXè siècle*
    (Paris, 1908), vol 1, 2nd edn, p. 59.

104 Thuillier, *Pour une histoire du quotidien*, 'La lessive', p. 129.

105 Perrot, p. 13.
106 Karl Marx, *Manuscrits* (edited in Paris in 1844), 3rd ms., p. 102. Quotation taken from Françoise Choay, *L'urbanisme. Utopies et réalités. Une anthologie* (Paris, 1965), p. 192.
107 Godin, *La richesse au service du peuple*, p. 32.

Chapter 3 The triumph of hygiene

1 Erwin H. Ackerknecht, *Medicine at the Paris Hospital, 1794-1848* (Baltimore, 1967), pp. 141ff.
2 The main publication in imperial Germany was the *Deutsche Vierteljahrschrifte für öffentliche Gesundheitspflege*, (German Public Health Quarterly), (1869–1914) which was published in Braunschweig. Almost every major city had a periodical dedicated to hygiene and health.
3 J.-P. Goubert, 'L'hygiène de l'eau dans la France contemporaine d'après les thèses de la Faculté de Médecine de Paris (1830–1940)', *Mensch und Gesundheit in der Geschichte*, (ed. Arthur Imhof) (Husum, 1980), pp. 59–78
4 *Bulletin de l'Académie de Médecine*, meeting of 6 October 1891.
5 Cartier (Dr), 'L'hygiène à Toulon', prix Vernois, 1895.
6 Meeting of the Académie de Médecine of 28 February 1893. Report by Netter, Thoinot and Proust.
7 *Bulletin de l'Académie de Médecine*, 24 April 1900.
8 F. Deligny, paper presented at the Paris conference, 1889.
9 J. Vidal, 'Du service des eaux alimentaires dans les campagnes', Paris Conference, 1889.
10 Alice Peeters, 'L'hygiène et les traditions de propreté', p. 23.
11 Mary Douglas, *Purity and danger; an analysis of concepts of pollution and taboo* (London, 1966).
12 Lepelletier de la Sarthe, *Voyage en Bretagne* (Paris, 1853).
13 Guillotin de Corson (abbé), *Les Pardons et les Pèlerinages de Basse-Bretagne* (Vannes, 1898).
14 Pierre Lazerges (Dr), 'Les Saints guérisseurs de Bretagne', unpublished medical thesis (Lyons, 1942).
15 By 14 November 1849, health commissions had been set up in forty-six departments (National Archives, F^8 169).
16 Decree issued by the mayor of Grand-Quevilly (Seine-Maritime) consulted in a file at the Archives of the Département du Nord, M 284.
17 Archives of the Département de l'Aude, 8 M 61, 2 May 1922.
18 Instructions received by the prefect of the Aisne, dated 2 June 1924; archives of the département de l'Aisne, 7 M 48.
19 In particular, the circulars of 10 December 1900 and 3 November 1902.
20 Circulars from the Ministry of Labour, Hygiene, Social Assistance and National Insurance dated 12 July 1924 and 24 June 1925. This text in part echoes a circular dating from 1904.
21 *Annuaire des eaux de la France pour 1851* (Paris, 1854), Introduction, p. XVIII.
22 Ibid., p. 48. For a similar description of the Bièvre in Paris and the

suburbs, cf. Parent-Duchatelet, *Essai sur les cloaques ou égouts*, pp. 70–1.

23 A. Franklin, *La Vie de Paris sous la Régence* (Paris, 1897), p. 263. N.B. This is a reprint of *Séjour de Paris* by Jean van Abcoude (Leyden, 1727).

24 *Annuaire des eaux de France pour 1851*, p. 83. Cf. also Jean-Pierre Chaline, 'Paysages industriels des vallées rouennaises', *Monuments historiques*, 122 (August–September, 1982), pp. 45–56.

25 Ed. Imbeaux (Dr), *Annuaire statistique et descriptif des distributions d'eau.*

26 *Revue britannique* (Brussels, 1850), vol. 1, p. 284.

Part II Mass Diffusion

Chapter 4 The power of the press

1 Gérard Lagneau, *Le Faire-Valoir. Une introduction à la sociologie des phénomènes publicitaires* (Paris, 1969), p. 70.

2 Ibid., p. 91.

3 Honoré de Balzac, *César Birotteau*, (1837), (Paris, 1975), p. 193.

4 Lagneau, *Le Faire-Valoir* p. 106.

5 Ibid., p. 23.

6 Ibid., p. 24.

7 The following sample was analysed: March 1843; December 1860; the years 1865, 1870, 1875, 1880, 1881, 1885, January–June 1890, July–December 1891; the years 1895, 1900, 1905, 1910, 1920, 1925, 1930, 1935; May–June 1936; the years 1940 and 1943.

8 Sample analysed: the years 1884, 1885, 1890, 1895, 1900, 1905, 1910, 1915, 1920, 1925, 1930, 1935, 1937.

9 Evelyne Diebolt, '*Le Petit Journal* et ses feuilletons, 1863–1914', thesis, (University of Paris VII, 1975), p. 13. Cf. also Marc Martin, 'La réussite du *Petit Journal* or the beginnings of the popular daily', *Bulletin du Centre d'histoire de la France contemporaine*, 3 (1982), pp. 11–36.

10 Michael B. Palmer, *Des petits journaux aus grandes agences. Naissance du journalisme moderne* (Paris, 1983), pp. 171–2 and 264–5; for circulation figures, cf. ibid., Annexes, pp. 331–41; cf. also A.-M. Durand and F. Labes, 'Une publication populaire originale: le supplément illustré du *Petit Journal*', doctoral thesis (Institut français de presse, Paris II, 1974).

11 Emile Durkheim, *Les formes élémentaires de la vie religieuse* (Paris, 1960), 4th edn, pp. 465ff.

12 Mircea Eliade, *La nostalgie des origines* (. . .) (Paris, 1971), pp. 222ff.

13 Auguste Comte, *Oeuvres choisies* (Paris, no date), 1st lesson, p. 59.

14 Luc Boltanski, 'Consommation médicale et rapport au corps', typescript (Paris, 1971), p. 83.

Chapter 5 The role of the hospital

1 Michel Foucault et al., *Les machines à guérir. (Aux origines de l'hôpital moderne)* (Paris, 1976), p. 18.

2 Ibid., p. 20.

3 Ibid., p. 20.

4 Marcel Candille 'Contribution à l'histoire de l'urbanisme de la capitale. La Collection des plans et dessins d'architecture de l'ancien Hôtel-Dieu de Paris', *Bulletin de la Société française d'histoire des hôpitaux*, 33 (1976), p. 86.

5 Françoise Hildesheimer, *Le Bureau de la santé de Marseille sous l'Ancien Régime* (. . .), (Marseilles, 1980), p. 20.

6 Foucault et al., *Les machines à guérir*, p. 136.

7 Ibid., p. 147.

8 This questionnaire, which is part of the Tenon Papers in the Bibliothèque Nationale (mss NAF 22742), was published by Foucault et al. in *Les machines à guérir*, pp. 158–67.

9 Muriel Jeorger, 'La structure hospitalière de la France sous l'Ancien Régime', *Annales ESC*, 5 (September–October, 1977), pp. 1033–4.

10 Archives of the department of Ille-et-Vilaine, C 1290, response to the 1724 survey.

11 Archives of the hospital of Malestoit, letter from the parish priest (23 July 1741), quoted in André Guillemot, 'L'hôpital de Malestoit du milieu du XVIIè siècle à la Révolution', masters rhesis, (1967), typescript, p. 207.

12 Ibid., p. 208.

13 Françoise Lelu, 'La population des "pauvres malades" de l'Hôtel-Dieu de Provins', master's dissertation (Paris I, 1976), typescript, p. 27.

14 Extract from the debate held at La Salpêtrière on 25 January 1756, quoted by Jacques Bernier, 'L'Hôpital de Bicêtre. 1763–1784', dissertation for the Ecole des Hautes Etudes en Sciences Sociales (Paris, 1975), typescript, pp. 257–8.

15 G. de Bertier de Sauvigny, *La France et les Français vus par les voyageurs américains, 1814–1848* (Paris, 1982), p. 158.

16 Isidore de Polinière, *Considérations sur la salubrité de l'Hôtel-Dieu et de la Charité* (Lyons, 1853).

17 National Archives, F[8] 199; extract from a report by Pluchart, General Inspector of Administrative Services, on the hospital at Croix (Nord).

18 Pierre Vallery Radot (Dr), *Nos hôpitaux parisiens. Un siècle d'histoire hospitalière (. . .) (1837–1947)* (Paris, 1948), p. 18.

19 M. Davenne, *Etudes sur les hôpitaux* (. . .) (Paris, 1862); M. Fosseyeux, *Les grands travaux hospitaliers de Paris au XIXè siècle* (Paris, 1912); C. Tollet, *Les édifices hospitaliers depuis leur origine jusqu'à nos jours* (Paris, 1892), 2nd edn.

20 Jules Guadet, *Eléments et théorie de l'architecture* (Paris, 1901), Book IX.

21 Bruno Foucart, 'Au paradis des hygiénistes', *Monuments historiques*, 114 (April–May 1981), p. 43.

22 Tollet, *Les édifices hospitaliers depuis leur origine*, foreword, p. VI.

23 Foucault et al., *Les machines à guérir*.

24 Tollet, *Les édifices hospitaliers depuis leur origine*, p. V.

25 P. Brouardel (Dean of the Faculty of Medicine of Paris), preface to ibid., p. VII.

26 Ibid., p. VIII.

27 Ibid., p. VIII.

28 Douglas, *Purity and Danger*.

29 Extract from the Report of the Academy of Sciences of 12 March 1788.

30 Clavereau was at the time the architect in the Department of Hospital Administration in Paris.

31 This programme also included recommendations and was called the *Blue Book*.

32 Meetings of 5, 19 and 26 October, 2, 9 and 23 November and 7 December 1864.

33 E. Trélat, *Etude critique sur la reconstruction de l'Hôtel-Dieu de Paris* (Paris, 1864).

34 Also called the 'Tollet system', it was approved by Larrey at the Academy of Sciences (meeting of 13 April 1874) and by Dr Hillairet at the Academy of Medicine (meeting of 16 March 1875).

35 They were thus described in order to distinguish them from the simpler, temporary hospital buildings (huts) also known as the 'American system'.

36 Tollet, *Les édifices hospitaliers depuis leur origine*, pp. 234–5.

37 Fosseyeux (former chief clerk and archivist of the National Assistance Board, doctor of letters), *Les grands travaux hospitaliers*, p. 39.

38 Robert Castel, *L'ordre psychiatrique. L'âge d'or de l'aliénisme* (Paris, 1976), p. 123.

39 M. Foucault, *Histoire de la folie à l'âge classique* (Paris, 1961), p. 381.

40 Text quoted by Ambroise Tardieu, *Dictionnaire d'Hygiène publique et de Salubrité*, vol. 2, p. 66.

41 *Le progrès médical*, 18 December 1937.

42 E. Bouchut and A. Després, *Dictionnaire de Médecine et de Thérapeutiques médicale et chirurgicale* (Paris, 1895), 6th edn, pp. 162–6. Cf. in particular Louis Fleury (Dr), *Traité thérapeutique et chimique d'hydrothérapie* (Paris, 1866), 3rd edn (1195 pages).

43 P.-H. Nysten, *Dictionnaire de Médecine* (Paris, 1858), 11th edn, p. 139.

44 A. Dechambre et al., *Dictionnaire usuel des sciences médicales* (Paris, no date), 3rd edn, p. 161.

45 P.-H. Nysten, *Dictionnaire de Médecine*, p. 139. The by-products in question were the extremities or the entrails of animals (head, feet, stomach, liver, gizzard).

46 Rostan, *Dictionnaire de Médecine*, 2nd edn, vol. 4, p. 542; Michel Lévy, *Traité d'hygiène publique et privé* (Paris, 1862), 4th edn, vol. 2, p. 159.

47 Tardieu, *Dictionnaire d'hygiène publique et de salubrité*, vol. 1, p. 184.

48 Lévy, *Traité d'hygiène publique et privé*, vol. 2, p. 156.

49 Ibid., pp. 155ff.

50 Nysten, *Dictionnaire de Médecine*, p. 161.

51 Dechambre et al., *Dictionnaire usuel des sciences médicales*, pp. 160–1.

52 Françoise Mayeur, *L'éducation des filles en France au XIXè siècle* (Paris, 1979), p. 44.

53 Lévy, *Traité d'hygiène publique et privé*, vol. 1, p. 181.

54 Ibid., p. 182.

55 Ibid., p. 180.

56 Françoise Loux, *Pratiques traditionnelles et Pratiques modernes d'hygiène et de prévention de la maladie chez les mères et leurs enfants* (Paris, 1975), p. 50.

57 Françoise Loux, *Le jeune enfant et son corps dans la médecine traditionnelle* (Paris, 1978), p. 190.

58 Ibid., p. 198.
59 Rémy de Gourmont, *Le chemin de velours* (Paris, 1911); text quoted in *Gérontologie*, 29 new series 1, (January 1979), p. 62.
60 Verdier, *Façons de dire, façons de faire*, p. 121.
61 Carmen Bernard, *Les vieux vont mourir à Nanterre* (Paris, 1978), p. 90.
62 Ibid., pp. 89–90.
63 National Archives, F^{15} 3651 – 3776 (1864 survey). Analysis of a sample of hospitals in twenty-seven departments, making up 37 per cent of the 1,405 hospitals in the France of that period.
64 National Archives, F^{15} 3732.
65 Ibid., F^{15} 3721.
66 Ibid., F^{15} 3657; the same applied to the Hôpital Saint Laurent in Langres (ibid., F^{15} 3725).
67 The total – of acceptances and refusals – does not add up to 100 per cent since 21 per cent of hospitals did not reply to this question in the 1864 survey.
68 National Archives, F^{15} 3740.
69 Ibid.
70 Three doctors failed to specify their sex!
71 The questionnaire was part of a survey financed by a research contract and directed by Emmanuel Le Roy Ladurie.
72 Reply given by a neuropsychiatrist, born in Paris in 1904.
73 Reply given by a doctor born in the department of the Rhône in 1901.
74 Reply given by a doctor born in Josselin (Morbihan) in 1900.
75 Reply given by a woman doctor born in Montguyon (Charente-Maritime) in 1903.
76 Reply given by a doctor born in Châteauroux (Indre) in 1903.
77 Reply given by a doctor born in Paris in about 1900.
78 Reply given by a Parisian doctor born about 1900.
79 Reply given by a doctor born in Châteauroux.
80 Between 53 and 60 per cent of them, depending on the question asked.

Chapter 6 Schools and the moulding of attitudes

1 The expression was coined by Jacques Léonard; cf. *Les médecins de l'Ouest au XIXè siècle* (Paris, 1978), 3 vols.
2 E. Aubert and A. Lapreste, *Cours élémentaire d'hygiène* (Paris, 1895), 2nd edn, preface, p. 5.
3 Cf. Elias, *La Civilisation des moeurs*, pp. 77ff.
4 If Octave Mirbeau is to be believed (cf. *Sébastien Roch* (Paris, 1973), pp. 77ff), supporters of the confessional school did not fail to offer 'certificates in hygiene' in their prospectuses.
5 Henri Napias (Dr), 'L'hygiène scolaire en France', *6th International Congress on Hygiene and Demography in Vienna – 1887*, Journal no. 12, p. 30.
6 C. Delvaille (Dr), extract from *La Gazette médicale* (Bayonne, 1880), p. 7.
7 Napias, 'L'hygiène scolaire en France' p. 37.
8 Ibid., p. 41.
9 Ibid., pp. 42ff.

10 Javal (Dr), *Rapport général fait au nom de la Commission d'hygiène scolaire* (Paris, 1884).

11 National Archives, F^{17} 7580.

12 Ibid., F^{17} 11 781, survey carried out in 1886–7 on medical inspection in primary schools.

13 Mangenot (Dr), 'De l'inspection hygiénique et médicale des écoles en France', *Revue d'hygiène* (June 1887).

14 Alexandre Layet (Dr), 'Enseignement de l'hygiène en France', *6th International Congress on Hygiene and Demography*, Journal no. 13, pp. 48–56.

15 *L'école et la famille. Journal d'éducation, d'instruction et de récréation*, 1 (1 January 1893).

16 Ibid., 10 (15 May 1899).

17 Ibid.

18 G. Gruno, *Le Tour de France par deux enfants. (Reading book for intermediate classes in primary schools)* (Paris, no date), p. 26.

19 L. Chabaud, *Lectures des Français* (Paris, 1887).

20 Jacques Ozouf, *Nous, les maîtres d'école* (Paris, 1967), p. 80.

21 Louis Boyer, *Le livre de morale des écoles primaires* (Paris, 1895), p. 125.

22 Jacques Defrance, 'Esquisse d'une histoire sociale de la gymnastique (1760–1870)', *Actes de la recherche en sciences sociales*, 6 (December 1976), pp. 22–46.

23 Cf. Michel Foucault, *Histoire de la sexualité (1). La volonté de savoir* (Paris, 1976), p. 39.

24 Chabaud, *Lectures des Français*, p. 10.

25 A. Léaud and E. Glay, *L'école primaire en France* (Paris, 1934), p. 184.

26 Ibid., p. 186.

27 Official instruction issued by the Ministry of Public Education (1887).

28 According to the survey quoted above, financed by a research contract between the Ecole des Hautes Etudes en Sciences Sociales and the DGRST (1976–7).

29 Léaud and Glay, *L'école primaire en France*, p. 188.

30 National Academy of Medicine, mss nos 204–8, Lycées impériaux (1867). Maxime Vernois, who was general inspector of the department of health and hygiene in secondary schools during the Second Empire, used this survey to draw up a *Code of Hygiene for Secondary Schools* (Paris, 1867).

31 The most isolated regions, particularly in central France, Brittany and mountainous regions.

32 Spring water: 11 lycées; river water: 6; well, tank and pond water: 6.

33 National Archives, F^{17} 7512, Central Committee for Hygiene, meeting of 18 November 1864.

34 Ibid., F^{17} 7512, *Rapports des proviseurs sur l'état sanitaire et la gestion économique des lycées*, Coutances, 30 November 1858.

35 Ibid., F^{17} 7586, *Etat sanitaire des lycées* (1868–9).

36 Ibid., Rapport sur l'hygiène du lycée de Lille, (no date). For the education of girls, cf. Françoise Mayeur, *L'enseignement secondaire de jeunes filles sous la IIIè Republique* (Paris, 1977); cf. also Guy Thuillier, 'Note sur les sources de l'histoire régionale de l'hygiène corporelle au XIXè siècle', *Revue d'Histoire Economique et Sociale*, 2 (1970), p. 233, n. 27.

37 G. Thuillier, 'Pour une histoire des gestes: en Nivernais au XIXè siècle', *Revue d'Histoire Economique et Sociale*, 2 (1973), p. 248, notes 120 and 121; and ibid; 'Pour une histoire régionale de l'eau (. . .)', Annales ESC, 1968, p. 67, n. 2.

38 Decrees of 10 July 1913 and 29 March 1914 on the installation of showers in factories. For secondary schools, cf. Méry (Dr) and Genevrier, *Hygiène scolaire* (Paris, 1914), p. 640, statistics by educational district for 1910.

39 National Archives, F^{17} 9784, survey of 24 April 1889 of boarding schools. Cf. also, ibid., F^{17} 11781, *Hygiène des ecoles primaires et des ecoles maternelles* (Paris, 1884).

40 These figures correspond to orders of magnitude and are indicative in nature. They are in fact based on a small sample; on the other hand, a schoolteacher's career usually began with a post in a rural area and finished with one in an urban area.

41 National Archives, F^{17} 14600, years 1935–9, primary education section.

42 See the case quoted by Jacques Ozouf, *Nous, les maîtres d'école*, p. 43; and the quotation from Claude Duneton, *Je suis comme une truie*, in the General Introduction, above.

43 Indeed, the development of an hygienic relationship to water also took place in private schools and various youth and popular education movements, particularly professional organizations, scouting and sports clubs. On these various points, see D.N. Baker and P.J. Harrigan, *The Making of Frenchmen* (. . .), 7, 2–3 (Summer/Autumn 1980), *Historical Reflections*; in this collection, cf. the articles by Raymond Labourie, Marc-Henry Soulet and Christian-Jean Guérin.

44 These editorials, published in *L'Aurore* of which Clémenceau was then editor-in-chief, have been analysed by Dr Michel Valentin; cf. *Histoire des sciences médicales*, VII, 3, (July–September 1973), pp. 245–52.

45 *L'Ouvrier vosgien. Journal d'éducation et de défense ouvrière*, 6th year, no. 269 (27 October 1907).

46 National Archives, F^{22} 543, Paris, 18 June 1907.

47 *L'Ouvrier vosgien*, 11 (July–August 1913), pp. 12–15.

48 This 'social alliance', dear to the heart of the school of Le Play, was distasteful to many; it claimed to bring together factory inspectors and workers' and employers' associations in the pursuit of hygiene, thus contributing to social peace and the happiness of all.

49 Based on the British model, this still very recent medical service in theory made quarterly visits to workers who did not dare to 'appear' before the doctor when they were not in a 'satisfactory' state of cleanliness.

50 For example, the *Journal de Santé*, an ephemeral periodical which appeared between 1789 and 1792; see also the items under the headings of 'Epidemics', 'Health' and 'Medicine in the home' that appeared in the weekly and daily press in the nineteenth century; cf. the *Journal des villes et des campagnes* during the July Monarchy; see also the amazing success of the *Manuel de Santé* edited by Raspail, and *La Santé universelle. Guide médical des familles*, by Dr Jules Massé, during the Second Empire.

51 For example, *Les Feuilles d'Hygiène, Le Bulletin des Feuilles d'Hygiène et de Philosophie* (1907), *L'Eau* (a monthly publication which popularized public

health) (1908), *Les Annales de Santé*; *Le Journal de l'Université des Annales*, founded in 1907, published discussions on hygiene from its first number onwards.

52 Cf. Jules Massé, *La Santé universelle* 'Popular hygiene course in 25 lessons' (Paris, 1854), pp. 193ff.

53 Guy de Maupassant, *Contes et Nouvelles* (Paris, 1974), vol. 1, 'Voyage de santé', pp. 82–7.

54 See, among other periodicals, *Les Annales des Ponts-et-Chaussées* (founded in 1831), *Les Annales du régime des eaux* (1887), *La Gazette des eaux* and *La Revue pratique d'hygiène municipale, urbaine et rurale* (1908).

55 Anne Querrien, *Généalogie des équipements de normalisation. L'école primaire* (Paris, 1975); and Querrien, 'L'enseignement. 1 – L'école primaire', *Recherches* 23 (June 1976).

56 Without forgetting the marked ethnocentricity that still persists today. Cf. J.M. Kalbermatten, S. Julius, Ch. G. Gunnerson, *Solutions appropriées des problèmes de développement: évaluation technique et économique* (Washington, 1979).

57 Yvonne Knibiehler, 'Les médecins et la "nature féminine" au temps du code civil', *Annales ESC*, 4 (July–August 1976), pp. 824–5.

Part III

Chapter 7 Water becomes an industrial product

1 Besnard, *Souvenirs d'un nonagénaire*, vol. 1, pp. 210–11.

2 Mercier, *Tableau de Paris*, vol. 2, ch. DCLXXI, pp. 339–41.

3 Nicolas Delamare, *Traité de police* (Paris, 1705), vol. 1, section 3, 'De la police de l'eau par rapport à la santé', pp. 544ff.

4 Thuillier, p. 52, n. 3.

5 Ibid., p. 56.

6 Jean Bouchary, *L'eau à Paris à la fin du XVIIIè siècle* (. . .) (Paris, 1946).

7 The *livre tournois* ('the livre minted at Tours'), which had been stable since 1726, had by the eve of the Revolution undergone only a slight devaluation of 5 per cent. Still defined as the same weight of gold, it became known as the franc in 1803 and remained stable until 1914 under the name of *franc-or* or 'gold franc'. Subsequently, it underwent a series of rapid devaluations.

Definition of the franc in terms of its weight in grams of gold

livre/*franc-or*	'Poincaré' franc	franc in 1939	franc in 1985
0.3	0.065	0.027	0.008

In view of the great changes that took place during the nineteenth century and the first half of the twentieth century, it is difficult to suggest anything other than approximate equivalences. If the average income of the manual worker is used as a basis, a coefficient of 150 has to be used in order to convert the *franc-or* into 1985 francs (1 *franc-or* = 150 francs).

8 I am grateful to Daniel Roche for having indicated to me the prices of water in Paris at the end of the eighteenth century.

9 Defer was offering a hogshead of water (238 litres) in perpetuity at 216 rather than 1,050 livres.

10 Eugène Belgrand, *Les Travaux souterrains de Paris* (. . .) (Paris, 1880), vol. 3, p. 356. Belgrand (1810–8) was a French engineer and author of works on the sanitation of Paris and its water supply. Independent member of the Académie des Sciences (1871).

11 Such as the Flachat company quoted by Guy Thuillier, op. cit., p. 55.

12 National Archives, F^{12} 6785, 14 December 1853.

13 Ibid., printed report, 19 pages.

14 A small minority of the shareholders of 1853 were foreign, mainly British, with some Italians.

15 The price of cast iron increased sharply between February and October 1853; this rise alone made a profit of 400,000 *francs-or* for the Campagnie Générale des Eaux.

16 Cf. Emile Brunneau, *Rapport à la commission municipale sur les traités intervenus entre M. le Conseiller d'Etat, administrateur du département du Rhône, la Compagnie générale des eaux de France et la Compagnie lyonnaise du gaz pour la distribution d'eaux potables et la continuation de l'éclairage à Lyon* (Lyons, 1853).

17 Compagnie Générale des Eaux, *Rapport du conseil d'administration (Report of the Board of Management) du 29 avril 1854.*

18 10 to 20 litres per inhabitant per day at the beginning of the nineteenth century; 30 to 50 litres around 1860; 100 to 120 litres around 1880; 3 to 400 around 1930.

19 According to the *Album du Centenaire* (Paris, 1953), published by the company, the price of the water sold by the Compagnie Générale des Eaux was always lower than the general wholesale price index between 1853 and 1952, with the exception of the period 1948–52.

20 On the subject of the Lyonnaise, cf. Jean Bouvier, *Le Crédit Lyonnais de 1863 à 1882* (Paris, 1961), vol. 1, pp. 488–93.

21 In 1973, the various management methods were distributed as follows: in 60 per cent of local authority districts, the supply network was managed by a third party with a financial stake in the operation, in 16 per cent by a lease holder, in 10 per cent by mixed methods and in 1 per cent by a company paid an agreed price to manage the network but without a direct financial interest in the operation. Figures from 'L'eau', *Après-demain*, 185 (June–August 1976).

22 G. Bechmann (chief engineer in the Highways Department), 'Enquête statistique sur l'hygiène urbaine dans les villes françaises', *Revue d'hygiène et de police sanitaire*, July–August 1892. A survey carried out in 1913 reached the same conclusions.

23 This practice was accepted and implemented in ten out of eighty-five departments. According to M. Barral, appendix II.

24 *Journal Officiel*, 28 and 30 October 1934, Minister of Agriculture circular

giving general instructions on the supply of drinking water at parish level.

25 Ibid., 9 November 1934, pp. 5–6.

26 Ibid., circular addressed to prefects.

27 Ibid., 9 December 1936, p. 12663.

28 Ibid., 5 March 1937, p. 2735.

29 Ibid., 7 May 1939, p. 5794.

30 See the *Rapport de la Cour des comptes (Audit Office Report)* (Paris, 1976) and the articles that appeared at this time in the daily and weekly press.

31 Initial expenditure is calculated on the basis of the following five elements: construction, maintenance and operating costs; the cost of any water treatment that may be necessary; energy costs (pumped or compressed water); depreciation; without counting various charges such as the installation and maintenance of meters.

32 Total expenditure was 6,398,593 francs for 593,142 inhabitants. On the value of the *franc-or* and its value in present-day francs, see note 7 above.

33 In fifty-one departments out of fifty-eight (88 per cent) there were water supplies taken from springs.

34 National Archives, F^8 184, no date. Not including the 2,600,000 francs spent in the parish of Boissey in order to supply Caen.

35 For example, at Brioude (Haute-Loire) in about 1895, according to Debauve, vol. 3, p. 275. The earthenware pipes installed between 1880 and 1900 accounted for only 4 per cent of the total, compared with the 78 per cent accounted for by cast-iron pipes. These figures are based on the projects for towns with more than 5,000 inhabitants submitted to the Higher Committee for Public Health in France (1884–1902).

36 Ibid., vol. 2, pp. 272–81.

37 This was particularly the case in two cities, Lyons and Villefranche-sur-Saône, supplied by the Compagnie Générale des Eaux.

38 The size of the local authority districts does not appear to have been a differentiating factor here.

39 The city of Lyons, with an initial investment cost of 17,000,000 francs, exerts disproportionate influence as far as the indirect management method is concerned; if Lyons were excluded, the average for this method of management would fall to 320,000 francs.

40 The basic price calculated in this way includes initial expenditure (cf. note 31 above) and costs for the year 1894, excluding profit margins. Thus in no case is it a question of the price paid by the consumer.

41 The sample has gaps (sic) which make it impossible to discover the basic price charged by companies and concession holders in nineteen out of thirty-three cases.

42 In 1976, the price of water in Paris was significantly lower (1.20 francs per cubic meter) when the supply was in public hands than when it was in private hands (2.82 francs). *Le Monde*, 6 December 1978.

43 From 0.08 francs to 9.80 francs per cubic meter. A. Bellon, 'La Distribution de l'eau, outil de la régionalisation industrielle', *Revue d'Economie Industrielle*, 1982, p. 92.

44 National Archives, F^{12} 6785, *Rapport du Conseil d'Administration de la Compagnie Générale des Eaux*, 29 April 1854.

45 Compagnie Générale des Eaux, *Album du Centenaire*, p. 45.
46 The price charged to industrial users was a function of their annual consumption: in order to obtain water at 0.30 francs, this consumption had to be greater than 8 cubic metres per day for a year.
47 Thuillier, p. 61, n. 2; price per cubic metre proposed in Nevers between 1880 and 1885 and imposed in 1910 by the use of meters.
48 Ibid., p. 61, n. 1.
49 Ibid., p. 60.
50 A. Wazon, p. 62.
51 16 francs for 3 people over 7 years of age and 16 francs (4 francs x 4) for 4 children under 7 (article 4 of the Paris Regulations of 1881).
52 Thuillier, p. 61.

Chapter 8 The development of the infrastructure

1 National Archives, C 838, Local loans.
2 Ibid.
3 Ibid., C 852.
4 That is to say analysis of the following files in the National Archives: F^{10} 2225 (Ain), 2228 (Aisne), 2254 (Ille-et-Vilaine, Indre, Indre-et-Loire).
5 National Archives, C 994 (1850).
6 Ibid., (1850).
7 Ibid., C 1047 (1856).
8 Ibid.
9 Ibid., C 5393.
10 National Archives, F^8 179–207 (1881–1902), analysis of a sample of twenty-five departments.
11 Bechmann, 'Enquête statistique sur l'hygiène urbaine', pp. 1062–9.
12 Unlike countries such as England, Belgium, Italy and Holland. Cf. E. Vallin (Dr), 'De l'étude et de l'exercice professionnel de l'hygiène', *Revue d'Hygiène et de Police sanitaire*, 1879, pp. 127ff.
13 Ibid., p. 137.
14 Four 5 franc tokens for four quarterly meetings.
15 'Notice sommaire sur l'état actuel de la distribution d'eau et de l'assainissement de Paris', *Annuaire sanitaire de France* (Paris, 1900), p. 1.
16 L. Thoinot (Dr), 'L'assainissement comparé de Paris et des grandes villes de l'Europe', *Annales d'hygiène publique et de médecine légale* (April 1898), pp. 289–331.
17 Waste water: liquid from cesspits and from the drainage tanks of some industrial establishments.
18 Gérard Jacquemet, 'Belleville au XIXè siècle', doctoral thesis (Paris, 1980), typescript; pp. 608–12 and 613–24 thesis published in 1984 under the same title.
19 Jacquemet, 'Urbanisme parisien: la bataille du tout-à-l'égout', pp. 505–48.
20 Edouard Baratier (ed.), *Histoire de Marseille* (Toulouse, 1973), pp. 333, 390, and 465.
21 François Lebrun (ed.), *Histoire d'Angers* (Toulouse, 1975), pp. 218–35.

22 Les Ponts-de-Cé were connected to the Angers water supply in 1924, Sainte-Gemmes in 1935, Trélazé in 1938.

23 Pierre Guillaume, *La Population de Bordeaux au XIXè siècle. Essai d'Histoire sociale* (Paris, 1972), p. 17.

24 Jean Meyer (ed.), *Histoire de Rennes* (Toulouse, 1972), pp. 396–8; cf. also J.-P. Goubert, 'Eaux publiques et démographie historique dans la France urbaine du XIXè siècle. Le cas de Rennes', *Annales de démographie historique*, 1975, pp. 115–22.

25 Thuillier, p. 57.

26 Ibid., p. 57, n. 6.

27 Ibid., p. 62, n. 2.

28 National Archives, F⁸ 220, report of 28 November 1912.

29 Alain Corbin, *Archaïsme et Modernité en Limousin au XIXè siècle, 1845–1880* (Paris, 1975), vol. 1, p. 81, n. 39.

30 On the case of Toulouse, cf. Mme Fidelle, 'Le Problème de l'eau à Toulouse et la consommation de 1945 à 1970', unpublished master's dissertation (Toulouse, no date). Cf. also Marie-Françoise Lormand, 'La santé publique à Toulouse de 1800 à 1870', unpublished master's dissertation (Toulouse, 1969).

31 Stendhal gave a brief description of this machine in his *Mémoires d'un touriste* in March 1838.

32 The same problem existed in Paris, where the (theoretical) output reached 180 litres per inhabitant per day in 1855, fell to 132 in 1874 and rose to only 146 in 1877. After Belgrand, *Les Travaux souterrains de Paris*, vol. 4.

33 Filtration galleries: subterranean aqueducts with a permeable bed which filters the flowing water.

34 In Toulouse, an 'inch' was equivalent to 13.89 litres.

35 P. Lavedan, *Histoire de l'urbanisme* (Paris, 1956), vol. 3, pp. 45ff.

36 Viré, *La Distribution publique d'eau à Bruxelles, 1830–1870.*

37 Blake, *Water for the Cities.*

38 G. Bechmann, *Salubrité urbaine. Distribution d'eau. Assainissement* (Paris, 1908).

39 Teissier-Rolland, *Etudes pour procurer de l'eau à la ville de Nîmes* (1843–4), 3 vols.

40 This sample (1884–1902) includes the following departments: Ain, Aisne, Alpes-Hautes, Alpes-Maritimes, Ariège, Aude, Bouches-du-Rhône, Calvados, Cantal, Doubs, Eure, Finistère, Garonne-Haute, Gironde, Ille-et-Vilaine, Isère, Marne-Haute, Maine-et-Loire, Mayenne, Nord, Puy-de-Dôme, Pyrénées-Basses, Rhône, Seine-et-Oise, Vendée.

41 In the sample used (a total of fifty-two parishes), the average population of the parishes with a wash-house was 2,342, while that of the parishes without a wash-house was 1,671.

42 Nineteen parishes out of 253 used these drainage systems; their average population was almost 7,000.

43 This applied mainly to rural parishes with between 500 and 1,500 inhabitants and a minority of urban districts with between 2,500 and 25,000 inhabitants.

44 The analysis in the book published by Henri Monod reached the same

conclusions: *L'Alimentation en eau potable* (. . .) (Melun, 1901). National Archives, F^8 180, 183, 191, 193, 199, 200. Survey carried out in 1892 on the order of the Minister of the Interior into drinking water supply schemes.

45 National Archives, F^{10} 2225, 26 February 1914.

46 Ibid., F^{10} 2254, Ille-et-Vilaine.

47 Ibid., 4 May 1939.

48 Thus Saint-Briac, in a period of rapid expansion, was connected in 1930 to the Dinard supply network (National Archives, F^{10} 2254).

Chapter 9 Water in everday life

1 Lise Vanderwielen, *Lise du plat-pays* (autobiographical novel written in the third person) (Lille, 1983), p. 60. At the time Lise was visiting her *petit bourgeois* aunt in Brussels.

2 Françoise Lehoux, *Le Cadre de vie des médecins parisiens aux XVIè et XVIIè siècles* (Paris, 1976), p. 1.

3 Information supplied by Micheline Baulant, to whom I am grateful.

4 Virole, 'Médecins et Chirurgiens à Paris dans la seconde moitié du XVIIIè siècle', p. 105.

5 Ibid., p. 194.

6 Léonard, *Les Médecins de l'Ouest au XIXè siècle*, doctoral thesis, Paris IV, typescript, 1976, vol. 6, pp. 1454–1460.

7 Results of the DGRST/EHESS survey (1977) on 'the hygiene of water'.

8 Memoirs of the Comtesse de Pange (née Pauline de Broglie), quoted by Leuilliot in his preface to Thuillier, *Pour une histoire du quotidien*, pp. XX–XXI.

9 Odile Arnold, *Le Corps et l'Ame. La vie des religieuses au XIXè siècle* (Paris, 1984), p. 81.

10 Statement by Mme B., in a letter from Louveciennes dated 3 June 1984.

11 *De la propreté, de son influence sur la santé, de ses effets sur le bien-être et la durée de l'existence* (Paris, 1827), p. 25. The medical school of Salerno, founded in the seventh century and famous in the twelfth century for its *Regimen Sanitatis Salernitanum*, written in verse, influenced treatises on cleanliness and civility until the nineteenth century.

12 Paul Sébillot, *Le Folklore de France. Les eaux douces* (Paris, 1983), new edn, p. 55.

13 Ibid., p. 60.

14 Jeanne Favret-Saada, *Les Mots, la Mort, les Sorts* (. . .) (Paris, 1977), p. 202, n. 8.

15 M. Bossi, *Statistique générale de la France. Département de l'Ain* (Paris, 1808), pp. 290–1.

16 Olivier Burgelin, 'Les outils de la toilette ou le contrôle des apparences', *Traverses*, 14/15 (April 1979), p. 37.

17 J. Léonard, *La France médicale au XIXè siècle* (Paris, 1978), pp. 204–5.

18 Corbin, vol. 1, pp. 80–1.

19 R.-H. Guerrand, 'L'imaginaire du logement social dans le roman français

au XIXè siècle', *Recherches*, 29 (December 1977), pp. 23–5.

20 Verdier, *Façons de dire, façons de faire*, pp. 108ff.

21 Thuillier, p. 52.

22 Mireille Laget, 'Naissance et Conscience de la vie: procréation, enfante-ment, obstétrique en Languedoc aux XVIIè et XVIIIè siècles', unpublished doctoral thesis (Paris IV, 1980), typescript, 3 vols. 'Babes in arms are changed only with regret. Its own filth hardens it and does it no harm' (p. 547).

23 Françoise Loux and Philippe Rochard, *Sagesses du corps. La santé et la maladie dans les proverbes français* (Paris, 1978).

24 Burguière, *Bretons de Plozévet*, pp. 156–62.

25 Verdier, *Façons de dire, façons de faire*, p. 115.

26 Riffault, 'De Chaptal à la mère Denis', pp. 257–63.

27 Imbeaux, *Annuaire statistique et descriptif des distributions d'eau*.

28 *Les Ouvriers des deux mondes* (Paris, 1892), 2nd series, vol. 3, p. 14.

29 Archives of the department of the Lower Rhine (Bas-Rhin), 5 M 325, decree issued by the mayor of Eywiller, 11 October 1860.

30 Timber work was replaced by iron supports, iron chains took the place of ropes, lips were raised and wells were covered over. These improvements, recorded by Guy Thuillier for the province of Nivernais in the nineteenth century, were also made in Ille-et-Vilaine and Aube (surveys in the O series of these two departments).

31 Thuillier, p. 50. Cf. also Lorgues, 'L'alimentation en eau des villages de Provence', pp. 33–41.

32 Underlined in the text.

33 Archives of the department of Meuse, 308 M1, 20 July 1908.

34 A reference to the great Public Health Act of 1902.

35 Archives of the department of Meuse, 308 M1, 20 July 1908.

36 Ibid., 14 September 1911.

37 Archives of the department of Deux-Sèvres, 6 M 18, report by the doctor in charge of epidemics, 12 July 1903.

38 Ibid., Poitiers, 28 February 1904.

39 Paquet (Dr) and F. Dienert, 'Rapport sur le contrôle des eaux d'alimentation du département de l'Oise', *Conseil Supérieur d'hygiène publique en France*, meeting of 29 March 1926 (Melun, 1926), 31 pages.

40 National Archives, F^8 215.

41 Archives of the department of Pas-de-Calais, M 2682, circular from Paris dated 18 July 1908 and signed by Georges Clemenceau, who was then Minister of the Interior.

42 National Archives, F^8 200, report by the Hygiene Committee, Royat, 2 April 1899.

43 Ibid., report by the Hygiene Committee, Clermont-Ferrand, 6 November 1900.

44 Ibid, F^8 201, favourable opinion expressed by the Consultative Committee on Public Health in France, 7 June 1886. The same was true in Mauran (Haute-Garonne) because of the cholera outbreak of 1884 (ibid., F^8 188 30 May 1893).

45 Ibid., F^8 179, opinion of the Consultative Committee on Public Health in France, 7 June 1886. The same was true in Le Faou (Finistère), ibid., F^8 188 (30 May 1893).

46 Marle, a municipal district of Laon in the department of Aisne; 2,456 inhabitants in 1896.

47 National Archives, F^8 180, report by Dr Blanquinque, 26 March 1900.

48 In Paris, Biarritz and Meudon (among other places), petitions were organized by the inhabitants between 1890 and 1900 in attempts to obtain a more plentiful supply of purer water.

49 Jeanne Favret, 'Le malheur biologique et sa répétition', Annales ESC, 1971, pp. 873–88.

50 National Archives, F^8 183, extract from the minute book of the municipal council of Jouques, 22 February 1900.

51 *Les Ouvriers des deux mondes*, pp. 12–14.

52 Ibid., pp. 57–8.

53 Ibid., pp. 76–104.

54 Each member of the family had a savings bank account. . . .

55 *Les Ouvriers des deux mondes*, p. 78.

56 Ibid., p. 97.

57 Alice Peeters, 'L'hygiène et les traditions de propreté'.

58 Jean Benoist, *Les Carnets d'un guérisseur réunionnais* (Saint-Denis, 1980), p. 33.

59 Simone Roux, *La Maison dans l'Histoire* (Paris, 1976), p. 230.

60 J.-P. Goubert, Culture technique, pp. 226–31.

61 Friedrich Engels, 'The Housing Question'.

62 See, for example, the ideal hygienic city described by Jules Verne in the *Cinq cents millions de la Bégum* (published in 1879).

63 Department of Internal Trade, Office of Sanitary and Industrial Regulations.

64 Archives of the departments of Haute-Marne (144 M1), Meurthe-et-Moselle (5 M 121), Dordogne (5 M 35), Charente-Maritime (7 M 8/1), Calvados (M 4083).

65 Pierrard, *La Vie ouvrière à Lille sous le Second Empire*, p. 92.

66 Henri Monod referred to this in 1888 in a brochure entitled 'Des pouvoirs en matière sanitaire' (Paris), p. 27. Roger-Henri Guenand, in his book, *Les origines du logement social en France* (Paris, 1969), makes the same point.

67 Archives of the department of Dordogne, 3 M 35 (1859).

68 Simone Roux, *La Maison dans l'Histoire*, p. 186.

69 Pierre Goubert, *Beauvais et le Beauvaisis de 1600 à 1730. Contribution à l'histoire sociale de la France du XVIIè siècle* (Paris, 1960), pp. 159–60.

70 Pierre Gouhier, *Le bâtiment* (. . .) (Paris-The Hague, 1971), vol. 1.

71 Pierre Goubert, *Beauvais et le Beauvaisis de 1600 à 1730*, pp. 229ff. Jean-Claude Perrot, *Genèse d'une ville moderne. Caen au XVIIIè siècle* (Paris-The Hague, 1975), vol. 1, pp. 92ff.

72 O. du Mesnil, *L'Habitation du pauvre* (Paris, 1890) (survey of the 13th *arrondissement* in Paris).

73 Pierrard, *La vie ouvrière à Lille sous le Second Empire*; cf. also Louis

Depouilly, 'L'Eau dans les logements ouvriers', medical thesis (Paris, 1900), n. 309.

74　According to the survey on the housing stock carried out in 1941–2 and published in Etudes et Conjonctures, article quoted, pp. 1183ff.

75　Quotations taken from; Lion Murard and Patrick Zylberman, 'Le Petit Travailleur infatigable' (. . .), *Recherches*, 25 (November 1976), p. 227.

76　Clerget, Clérice, Deroy et al., *Les Merveilles de l'Exposition de 1889* (Paris, no date), p. 566.

77　A. Quantin, *L'Exposition du Siècle, 14 April–12 November 1900* (Paris, no date), p. 291.

78　Riffault, 'De Chaptal à la mère Denis', p. 260.

79　These changes are very clearly highlighted in some very detailed, rather long-winded monographs, such as those devoted to Minot (CNRS survey) and Plozévet (DGRST survey).

80　*Les Ouvriers des deux mondes*, pp. 325–68, case 70 studied by P. du Marroussem.

81　Ibid., vol. IV (Paris, 1863), pp. 241–82.

82　Ibid., pp. 245–6.

83　R.-H. Guerrand, 'Petites histoires du quotidien: l'avènement de la chasse d'eau', *L'Histoire*, 43 (March 1982), pp. 96–9.

84　*Le Parisien chez lui au XIXè siècle, 1814–1914* (exhibition catalogue) (Paris, 1976). Cf. also R. Burnand, *La Vie quotidienne en France, 1870–1900* (Paris 1947).

85　National Archives, C 943, questions 5 and 9, hygiene conditions at work and in dwellings and nutrition of apprentices.

86　Ibid., Paris, 9 June 1848.

87　Ibid.

88　Ibid.

89　Ibid., C 3018, questionnaires A and B (1873).

90　Ibid., Sénones (Vosges).

91　*Le Musée social, mémoires et documents* (Paris, 1909), pp. 1-4, text quoted in *Milieux*, 5 (February 1981), p. 75.

92　National Archives, F^{22} 568, Paris, 24 October 1921, memorandum for the 3rd office of the Ministry of Labour and National Insurance.

93　Ibid., Paris, 25 May 1911.

94　Ibid., 20 January 1923.

95　Ibid., F^{22} 543, report by Louis Lisbonis, Marseilles, 1908.

96　Ibid., F^{22} 517, decree issued by the Ministry of Labour in pursuance of the law of 25 October 1919. Cf. also M. Ducros, 'Le saturnisme dans la typographie', *Bulletin de l'Inspection du Travail et de l'Hygiene Industrielle* (Paris, 1906), pp. 421–60.

97　Ibid., Survey of the cases of lead poisoning observed in Paris hospitals during the first three quarters of 1911, continued until 1921.

98　Ibid., F^{22} 512, Lille, 18 May 1914.

99　Decree issued by the Ministry of Labour in pursuance of the law of 25 October 1919 on occupational diseases, which came into force on 27 January 1921.

100 See *L'Hygiène Ouvrière*, a quarterly publication founded in 1910 as the organ of the Association for the Hygiene and Safety of Workers and the WorkPlace, which was set up in 1905.

101 National Archives, F^{22} 512, extract from reports of debates in the Senate, meeting of 9 February 1909.

102 Ibid., reply by Viviani, Minister of Labour.

103 Alexandre Girard (chemical engineer), 'La réglementation de l'emploi des fosses septiques', *Revue mensuelle de la Chambre syndicale des entrepreneurs de maçonnerie, ciments et béton armé de la ville de Paris* (. . .), May 1925, p. 61.

104 National Archives, F^{22} 513, letter to A. Weill (15, rue Scheffer, Paris, XVIè) to the chairman of the Committee on Hygiene and Salubrity, 2 September 1912.

105 Jeanne Gaillard, 'Paris, la Ville (1852–1870)', doctoral thesis (Lille-Paris, 1976), p. 208ff. These overcrowded conditions also existed in cities in the seventeenth and eighteenth centuries. Cf. Jean-Pierre Bardet, *Rouen aux XVIIè et XVIIIè siècles. Les mutations d'un espace social* (Paris, 1983), vol. 1, pp. 93ff.

106 National Archives, F^{22} 513, Paris, 13 May 1910.

107 Ibid., report of 18 May 1910.

108 Ibid., report by the departmental inspector, Bordeaux, 8 October 1905.

109 Ibid., same report.

110 Ibid., F^{22} 538, Paris, 15 February 1917.

111 Jules Michelet, *Journal* (Paris, 1976), vol. 3 (1861–7), 29 May 1867.

112 P. Guiral and G. Thuillier, *La Vie quotidienne des domestiques en France au XIXè siècle* (Paris, 1978), pp. 43–4.

113 H. de Balzac, *Les Paysans* (Paris, 1855), p. 44.

114 Jules Romains, *Le drapeau noir* (Paris, 1937).

115 Lawrence Wright, *Clean and Decent* (. . .) (London, 1960), p. 138.

116 Ibid., pp. 138–9.

117 Vincenz Priessnitz (1799–1851), a Silesian peasant farmer and proponent of hydrotherapy after 1830.

118 S. Giedion, *La mécanisation au pouvoir* (Paris, 1980), p. 538.

119 Jules Huret, *En Amérique. De New York à la Nouvelle-Orléans* (Paris, 1905), p. 22.

120 Victor Gelu, *Marseille au XIXè siècle* (Paris, 1972), p. 231.

121 Lucien Gaillard, *La Vie quotidienne des ouvriers provençaux au XIXè siècle* (Paris, 1981), p. 42.

122 Sébillot, *Le Folklore de France, les eaux douces* (Paris, 1983), p. 89.

123 Ibid., pp. 92–3.

124 Marc Leproux, *Dévotion et saints guérisseurs* (Paris, 1957), pp. 216–17.

125 Sébillot, *Le Folklore de France, les eaux douces*, p. 217.

126 Mgr Gaume, *L'Eau bénite au XIXè siècle* (Paris, 1866).

127 M. Leproux, *Dévotion et saints guériseurs*', p. 48.

128 P. Brouardel (Dr), 'Mortalité par fièvre typhoïde en France', *Recueil des travaux du Comité consultatif d'hygiène publique de France* (. . .), 1889, vol. 20 (extract).

129 *Statistique générale de la France*, new series, VII (1927), p. XII.

130 Charles Nicolle (Dr), *Destin des maladies infectieuses* (Paris, 1935), p. 265.
131 P. Marot (Dr), *Pathologie régionale de la France* (Paris, 1958), vol. I, pp. 66ff.
132 Ariège, Deux-Sèvres, Dordogne, Aude, Nord (according to my investigations in the archives of these departments).
133 Nicolle, *Destin des maladies infectieuses*, p. 265.
134 Brouardel, 'Mortalité par fièvre typhoïde en France', p. 17.
135 F. Chapuis, *Manuel de préparation militaire en France* (Paris–Nancy, 1914), pp. 210–11.
136 Brouardel, 'Mortalité per fièvre typhoïde en France', p. 16.
137 *Annuaire sanitaire de la France* (Paris, 1902), 1st edn, 'Résultats des travaux d'amenée d'eau sur la mortalité des communes de 1890 à 1898 inclus', pp. 357–85.
138 Brouardel, 'Mortalité per fièvre typhoïde en France', pp. 20–1.
139 Burnand, *La Vie quotidienne en France de 1870 à 1900* (Paris, 1947), p. 116.
140 Archives of the department of the Bas-Rhin (Lower Rhine), 5 M 235.
141 Grimmer, *Vivre à Aurillac au XVIIIè siècle*, pp. 193ff.
142 Account given by Jean Nicolas (according to the memoirs of a contemporary).
143 Quantin, *L'Exposition du siècle*, p. 211.
144 The expression comes from the German poet Novalis.
145 Bannour, *Eugénie de Guérin*, p. 31.
146 Ibid., p. 204.
147 Emile Zola, *Germinal* (1885) (Harmondsworth, Middx, 1954), pp. 118–20.
148 Thuillier, p. 67.
149 This 'cultural digestion' has been highlighted in particular by Michel de Certeau in *L'Invention du quotidien* (Paris, 1980), vol. 1, pp. 10ff.

Conclusion: the Significance of Change

1 Annie Quartararo-Vinas, 'Médecins et Médecine dans les Hautes-Pyrénées au XIXè siècle', Tarbes, 1982, p. 227.
2 For example, the sociability of the traditional wash, the well, the washhouse and the fountain.
3 A structural theory of the same type has been vigorously expressed by Jacques Attali in *L'Ordre cannibale* (Paris, 1979), pp. 13ff.
4 I recall the jealously guarded secrecy surrounding the financial records of the large private companies and the silence of the people questioned about the price of water.
5 Jean Lorcin, 'Le projet du Lignon: une tentative d'application du "socialisme municipal" à Saint-Etienne en 1900', *Bulletin du Centre d'Histoire économique et sociale de la région lyonnaise*, 4 (1973), pp. 63–76.
6 Jacques Léonard, 'Les Médecins de l'Ouest au XIXè siècle', vol. 5, pp. 1160–2, typescript.
7 Serge Moscovici, *Essai sur l'histoire humaine de la nature* (Paris, 1977), p. 21.

Further Reading

Please note: The archive sources indicated in the notes are not listed here

The development of the infrastructure

France

Compagnie générale des eaux, *Album du centenaire* Paris, 1953. *Annuaire sanitaire de France*, Paris, 1900.

Barral, M., *De l'alimentation en eau potable des communes*, Paris, 1914.

Barraqué, B., *Approches d'Environnement et Sociétés locales* (. . .), Ministère de l'Environnement, Ministère de l'Urbanisme et du Logement, Paris, 1985.

Bechmann, G., *Salubrité urbaine. Distribution d'eau, assainissement*, Paris, 1908.

Belgrand, E., *Monographie des eaux de Paris*, Paris, 1875.

Belgrand, E., *Les Travaux souterrains de Paris*, vols. 3 and 4, Paris, 1882.

Bonnin, J., *l'Eau dans l'antiquité* (. . .), Paris, 1984.

Bouchary, j., *l'Eau à Paris à la fin due XVIIIè siècle* (. . .), Paris, 1946.

Claude, V., 'Strasbourg 1850–1914. Assainissement et politiques urbaines', Thesis, Ecole des Hautes Etudes en Sciences Sociales, 1985, typescript.

Debauwe, A., *Distributions d'eau. Egouts*, Paris, 1897, 2 vols.

Gay, J., *L'Amélioration de l'existence à Paris sous le règne de Napolèon III* (. . .), Geneva, 1986.

Gille, B., 'L'eau', *Histoire générale des techniques*, vol. 2, Paris, 1964.

Guillerme, A., *Les Temps de l'eau* (. . .), Seyssel, 1983.

Imbeaux, Ed., *Annuaire statistique et descriptif des distributions de l'eau*, Paris, 1909, 2nd edn.

Lacordaire, S., *Les Inconnus de la Seine*, Paris, 1985.

Léonard, J., *Archives du corps. La santé au XIXè siècle*, Rennes, 1986.

Mille, A., *Assainissement des villes par l'eau, les égouts, les irrigations*, Paris, 1886.

Monod, H., *L,Alimentation publique en eau potable de 1890 à 1897*, Melun, 1901.

Trélat, E., *La Salubrité*, Paris, 1899.

England

Briggs, A., *Victorian Cities*, London, 1968.
Dickinson, H. W., *Water Supply of Greater London*, Leamington Spa and London, 1954.
Robins, F. W., *The Story of Water Supply*, Oxford, 1946.
Saunders, P., *The Regional State: a Review of the Literature and Agenda for Research*, University of Sussex, 1983.

Italy

Bortoletti, L. Da., *Storia della politica edilizia in Italia*, Roma, 1978.
Calabi, D., 'I servizi tecnici a rete e la questione della municipalizzazione nelle città italiane (1880–1910)' in Morachiello (P.) et alii, *Le Macchine imperfette*, Officina Edizioni, 1979.

Belgium

Viré, L., *La distribution publique d'eau à Bruxelles 1830–1870*, Bruxelles, 1073.
Grahn, E., *Die städtische Wasserversorgung im Deutschen Reiche*, vol. 2, München und Berlin, 1902.
Grahn, E., 'Die deutschen Städte', *Die städischen Wasserwerke*, Wuttke, R. (Hrsgb.), vol. 1, Leipzig, 1904.
Winkler, H., 'Wasserversorgung und Abwasserbeseitigung als Probleme der Bielefelder Stadtpolitik in der zweiten Hälfte des 19. Jahrhunderts', Universität Bielefeld, Fakultät für Geschichtswissenschaft, 1986, typescript.

USA

Blake, N. M., *Water for the Cities. A History of the Urban Water Supply Problem in the United States*, Syracuse (USA), 1956.

Switzerland

Sardet, F., 'Ode inachevée à la boue. Hygiène publique à Genèva (1702–1792)', master's dissertation, University of Paris (Panthéon-Sorbonne), 1984, typescript.

Algeria

Arrus, R., *L'Eau en Algérie de l'impérialisme au développement (1830–1962)*, Alger-Grenoble, 1985.

Water, hygiene and school

France

Aubert. E. et Lapresté A., *Cours élémentaire d'hygiène*, Paris, 1895, 2nd edn.
Bert, P., *L'Instruction civique à l'école* (. . .), Paris, 1882.

Boyer, L., *La Livre de morale des écoles primaires*, Paris, 1895.

Bruno, G., *La Tour de la France par deux enfants, Livre de lecture courante. Cours moyen*, Paris, no date.

Federici, P., 'L'Hygiène dans l'enseignement primaire au XIXè siècle', medical thesis, Paris VI, 1980, typescript.

Gerbod, P., *La Vie quotidienne dans les lycées et collèges au XIXè siècle*, Paris, 1968.

Javal, Dr, *Rapport général fait au nom de la commission d'hygiène scolaire*, Paris, 1884.

Méry Dr, et Genevrier, *Hygiène scolaire*, Paris, 1914.

Ozouf, J., *Nous les maîtres d'école* (. . .), Paris, 1977.

English-speaking countries

Sutherland, N., 'To create a strong and healthy race: school children in the Public Health Movement, 1830–1914', *History of Education Quarterly*, XII, 3, (Fall 1971).

Italy

Vivante, R., *La malattie trasmissibili e il buon governo della Scuola ad uso dei maestri elementari*, Venice, 1934.

Water and housing

France

Cheysson, E., 'La question des habitations ouvrières en France et à l'Etranger', *Revue d'Hygiène*, August 1886.

Depouilly, L., 'L'Eau dans les logements ouvriers', medical thesis, Paris, 1900, n. 309.

Eleb-Vidal, M. and Debarre-Blanchard E., 'Architecture domestique et mentalités. Les traités et les pratiques, XVIè – XIXè siècles', *In Extenso*, 2, 1984.

Fonssagrives Dr, *Etude d'hygiène et de bien-être domestiques*, Paris, 1870.

Guerrand, R.-H., *Les Origines du logement social en France*, Paris, 1966.

Mesnil, O. du, *L'Habitation du pauvre*, Paris, 1890.

Napias, H. and Martin A. J., *L'Etude et les progrès de L'hygiene en France de 1878 à 1882*, Paris, 1882.

Roux, S., *La Maison dans l'Histoire*, Paris, 1976.

Teyssot, G., 'H.B.M.: expérience et pauvreté, *Urbi*, 5, 1982.

Germany

Artelt, et al., eds, *Städte-, Wohnungs- und Kleider- hygiene des 19. Jahrhunderts*, Stuttgart, 1969.

Engels, F., *La Question du logement*, Paris, 1969 (new edn).

Italy

Alberti, S., 'The housing Problem in Italy', *Quarterly Review*, 7 October 1948.
Chimenti, C., 'La casa indossata: le culture dell' abitare nell' Italia del Dopoguerra', *Parametro*, 127, 1984.
Erba, A., 'La casa popolare nei suoi vari aspetti storici igienico-sanitari ed etico-sociali', *Edilizia popolare*, 5, 1955.

USA

Giedion, S., *Mechanization takes command*, New York, 1948.

Water and hospitals

France

Davenne, M., *Etudes sur les hôpitaux* (. . .), Paris, 1862.
Fosseyeux, M., *Les Grands Travaux hospitaliers de Paris au XIXè siècle*, Paris, 1912.
Foucault, M. et al., *Les Machines à guérir* (. . .), Paris, 1976.
Tollet, C., *Les Edifices hospitaliers depuis leur origine jusqu'à nos jours*, Paris, 1892.

The social uses of water

Allen, E., *Wash and Brush up*, London, 1976.
Bachelard, G., *L'Eau et les Rêves. Essai sur l'imagination de la matière*, Paris, 1942.
Bologne, J.-Cl., *Histoire de la pudeur*, Paris, 1986.
Buffet, B. et Evrard, R., *L'Eau potable à travers les âges*, Liège, 1950.
Cabanès, Dr, *Moeurs intimes du passé (2nd series). La vie aux bains*, Paris, 1954.
Corbin, A., *Le Miasme et la Jonquille. L'odorat et l'imaginaire social. XVII–XVIIIè siècle*, Paris, 1982.
Csergo, J., *Liberté, égalité, propreté. La morale de l'hygiène au XIXè siècle*, Paris, 1988.
Csergo, J., 'L'hygiène à domicile. Eau et propreté corporelle à Paris. 1850–1900', *Le Corps et la Santé*, 1, Paris, 1985.
Douglas, M., *De la souillure. Essais sur les notions de pollution et de tabou*, Paris, 1971.
Durkheim, E., *Les Forme élémentaires de la vie religieuse*, Paris, 1960, 4th edn.
Eliade, M., *La Nostalgie des origines* (. . .), Paris, 1971.
Franklin, M., *La Vie privée d'autrefois: l'hygiène*, Paris, 1890.
Girard, L., *La Politique des travaux publiques sous le Second Empire*, Paris, 1952.
Gleichmann P.-R., 'Die Verhäuslichung von Harn- und Kotentleerungen', *Mensch, Gesellschaft*, 4, 1977, pp. 46–52.
Gleichmann, P.-R., 'Städte reinigen und geruchlos machen, menschliche Körperentleerungen, ihre Geräte und ihre Verhäuslichung', in H. Sturm

(ed.), *Ästhetik und Umwelt*, Tübingen, 1979.

Guerrand R.-H., *Les Lieux. Histoire des commodités*, Paris 1985.

Heller, G., *Propre en ordre. Habitation et vie domestique. 1850–1930: l'exemple vaudois*, Lausanne, 1979.

Lambton, L., *Temples of Convenience*, London, 1978

Laporte, D., *Histoire de la merde*, Paris, 1978.

Loux, F., *Le Corps dans la sociéte traditionnelle*, Paris, 1979.

Murard, L. and Zylberman, P., 'L'haleine des faubougs: ville, habitat et santé au XIXè siècle', *Recherches*, 29, 1977.

Netter, A. and Rozenbaum, H., *Histoire illustrée de la contraception (. . .)*, Paris, 1985

Reynolds, R., *Cleanliness and Godliness, or the Further Metamorphosis*, London, 1943.

Rival, N., *Histoire anecdotique de la propreté et des soins corporels*, Paris, 1985.

Sébillot, P., *Le Folklore de France. La mer et les eaux douces*, Paris, 1905

Thuillier, G., *Pour une histoire du quotidien au XIXè siècle en Nivernais*, Paris–The Hague, 1977.

Verdier, Y., *Façons de dire, façons de faire. La laveuse, la couturière, la cuisinière*, Paris, 1979.

Vigarello, G., *Le Propre et le Sale. L'hygiène du corps depuis le Moyen Age*, Paris, 1985.

Wright, L., *Clean and Decent. The Fascinating History of the Bathroom and the Water Closet (. . .)*, London, 1960.

Water, health and disease

France

Bourdelais, P. and Dodin, A., *Visages du choléra*, Paris, 1987.

Bourdelais, P. and Raulot J.-Y., *Une peur bleue. Histoire du choléra en France, 1832–1854*, Paris, 1987

Bouteiller, M., *Médecine populaire d'hier et d'aujourd'hui*, Paris, 1966.

Boutin, P., 'Points de repère pour une histoire de l'assainissement', *CEMAGREF, BI* 314–15 (March–April 1984), pp. 41–9.

Boutin, P., 'Eléments pour une histoire des procédés de traitement des eaux résiduaires', *La Tribune Cebedeau*, 511–12, pp. 3–18.

Brouardel, P. and Thoinot, L., *La Fièvre typhoïde*, Paris, 1895.

Cabanès, Dr, *Remèdes d'autrefois*, Paris, 1905.

Dubos, R., *Mirage de la santé*, Paris, 1961.

Dupuy, G. and Knaebel, G., *Assainir la ville hier et aujourd'hui*, Paris, 1982.

Jacquemet, G., 'Urbanisme parisien: la bataille du tout-à-l'égoût à la fin du XIXè siècle', *Revue d'Historie moderne et contemporaine*, XXVI (Oct–Dec 1979), pp. 505–48.

Léonard, J., *Les Médecins de l'Ouest au XIXè siècle*, Paris–Lille, 1978 (3 vols).

Leproux, M., *Dévotions et Saints guérisseurs*, Paris, 1957.

Marot, R., *Pathologie régionale de la France*, National Institute of Hygiene monograph, n. 16 and 17, Paris, 1958, 2 vols.

Michel, Dr, *De l'influence de l'eau potable sur la santé publique*, Paris, 1884.
Nicolle, Ch., *Destin des maladies infectieuses*, Paris, 1933.
Richet, Ch., *Pathologie de la misère*, Paris, 1957.
Sorre, Max., *Les Fondements biologiques de la géographie humaine* (. . .), Paris, 1943.

England

Chadwick, E., *Report on the sanitary conditions of the labouring population of Great Britain*, B.P.P., 1842 (H.L.), vol. XXVI.
McKeown, T., *The Role of Medicine*, Oxford, 1976.
Tristan, F., *Promenades dans Londres* (. . .), ed. F. Bédarida, Paris, 1978.
Woods, R. and Woodward J., (eds), *Urban Disease and Mortality in Nineteenth-Century England*, London and New York, 1984.

Germany

Spree, R., *Soziale Ungleicheit vor Krankheit und Tod. Zur Sozialgeschichte des Gesundheitsbereichs im Deutschen Kaiserreich*, Göttingen, 1981.

Canada

Defries, R. D. (ed.), *The Development of Public Health in Canada*, Canadian Public Health Association, 1940.
Farley, M., Keel, O. and Limoges, C., 'Les commencements de l'administration montréalaise de la santé publique (1865–1885)', *Journal of the History of Canadian Science, Technology and Medicine/Revue d'Histoire des sciences, des techniques et de la médecine au Canada*, 20, pp. 24–46; 21, pp. 85–109, 1982.
Roland, C. (ed.), *Health, Diseases and Medicine. Essays in Canadian History*, Toronto, 1983.

USA

Meeker, E., 'The improving Health of the United States, 1850–1915,' *Explorations in Economic History*, 9, 4 (Summer 1092), pp. 353–74.
Weiss, H. and Kemble, H., *The Great American Water-cure Craze*, Treton, 1967.

For a more extensive bibliography, see:

Goubert, J. P., 'La Conquête de l'eau. Analyse historique du rapport à l'eau dans la France contemporaine (1830–1940)', doctoral thesis (History) University of Paris-VII, 1984, 3 vols (typescript).

Some Special issues of Journals (in France)

Congrès National des Sociétés savantes (111è), *Usages et Représentations de l'eau*, Poitiers, 1986.
Corps écrit, 'L'eau', Paris, 1985.

Etudes Rurales, 'Les Hommes et l'eau', 93–4, 1984.

Le Monde alpin et rhodanien, 'Usages et images de l'eau', 4, 1985.

Les Annales de la Recherche urbaine, 'L'eau dans la ville' April, 30, 1986.

L'Outre-Forêt, Revue d'Histoire de l'Alsace du Nord, 'Au fil de l'eau', 57, 1987.

Mémoire d'Ardèche. Temps présent, 'Ça coule de source. L'eau et ses usagers en Ardèche', 11–12, 1986.

Monuments Historiques, 'L'eau douce', 122, 1982.

Total Information, 'L'eau', 104, 1986.

Géo, 'L'eau – Source de vie en danger', 112, June 1988.

Index

Index by Isobel McLean